T0213535

Basic Organic Chemistry for the Life Sciences

Hrvoj Vančik

Basic Organic Chemistry
for the Life Sciences

Second Edition

Hrvoj Vančik
Department of Chemistry
University of Zagreb
Zagreb, Croatia

ISBN 978-3-030-92440-9   ISBN 978-3-030-92438-6   (eBook)
https://doi.org/10.1007/978-3-030-92438-6

The first edition of this book has been the result of my collaboration with plenty of my colleagues and students during more than twenty years of my lecturing of basic organic chemistry. The manuscript of the first edition of this textbook was not be possible without the insightful comments of reviewers Mladen Mintas, Miroslav Bajić, Srđanka Tomić-Pisarović, (University of Zagreb, Croatia) and Igor Novak (Charles Sturt University, Sidney, Australia), to whom I owe a debt of gratitude. I also thank my colleagues Zlatko Mihalić and Đani Škalamera (University of Zagreb, Croatia) for their fruitful comments in the preparation of this second edition.

Preface to the Second Edition

The first edition of this textbook has appeared as a result of the experience in more than twenty years of lecturing organic chemistry to students of biology, molecular biology, and ecology within the Faculty of Science and Mathematics at the University of Zagreb. Since the great books of organic chemistry for chemists appear to be too advanced for students whose study is only partially related to chemistry, I have been decided to prepare the text that is more oriented to the essence of organic chemistry.

Open problems in writing the basic organic chemistry textbook include not only the selection of the concepts for the representation of the material, but also the level of the explanation of the complex phenomena such as reaction mechanisms or electron structure. Here I propose the compromises. First compromise is related to the mode of the systematization of the contents, which can traditionally be based either on the classes of compounds, or on the classes of reactions. Here, the main chapter titles contain the reaction types, but the subtitles involve the compound classes. The electronic effects as well as the nature of the chemical bond is described by using the quasi-classical approach starting with the wave nature of the electron and building the molecular orbitals from the linear combination of the atomic orbitals on the principle of the qualitative MO model. Hybridization is avoided because all the phenomena on this level can be simply explained by non-hybridized molecular orbitals.

The text is divided into two parts. First chapters deal with fundamental aspects of the structural theory, reaction dynamics of organic reactions, electronic structure, and some basic spectroscopy. In this, the second edition, I have extended the section about pericyclic reactions. The new section about the principles of organic synthesis is also added. Last, the largest chapter represents the introduction to the organic chemistry of natural products. Comparison of the reactions in the laboratory with the analogous molecular transformations in living cells will help the students to understand the basic principles of biochemistry. The most interesting property of organic chemical systems, the formation of the high diversity of structures, is pointed out almost in all chapters. This approach is designed to help the students to provide deeper insight into the phenomena of the chemical evolution as a base for the biological evolution.

At the end of this edition, the chapter about molecular receptors, supermolecular and supramolecular chemistry, as well as about the general principles and the mechanisms of enzyme catalysis is added.

I intend this book for students of biology, molecular biology, ecology, medicine, agriculture, forestry, and other professions where the knowledge of organic chemistry plays the important role. The content is designed on the way to be helpful for the readers in their further study of biochemistry. I also hope that the work could also be of interest to non-professionals, as well as to the high school teachers.

Zagreb, Croatia   Hrvoj Vančik
September 2021

Introduction

The historic moment when organic chemistry which deals with compounds originating from living organisms appeared as a subdiscipline of chemistry is difficult to establish. More than two hundred years ago, in 1784, T. Bergman used the term *organic chemistry* for the first time. Perhaps, two historic advancements in the development of chemistry could be regarded as crucial for the development of this branch of science.

First are the investigations of **Jöns Jacob Berzelius** who at the beginning of the XIX century has developed the method for systematic elemental analysis of organic substances. Berzelius has observed that, after burning, all organic substances produce carbon dioxide and water. By accurately measuring the masses of these products, he has calculated the percentages of carbon, hydrogen, and oxygen in organic compounds. The most important conclusion was essential for the development of organic chemistry: All organic compounds consist of carbon and hydrogen. Accordingly, organic chemistry could also be called the chemistry of carbon compounds.

Secondly, perhaps the most important event in the development of organic chemistry was the discovery that organic compounds could be prepared from inorganic precursors. In 1828, **Friedrich Wöhler** has successfully prepared the organic compound urea simply by heating the inorganic salt ammonium cyanate. Before this discovery, the origins of organic substances were thought to be exclusive in living organisms.

One of the fundamental and general questions about the nature of organic compounds is the special nature of carbon as the basic element from which all the organic substances and all the known substances of life are built. One of the basic questions in chemistry is: Why, for instance, silicon, which is together with carbon in the same group of periodic table, is not the element for the development of an alternative version of life?

To answer this question, it is necessary to take a look on the special properties of the element carbon, the properties which are responsible for the appearance and evolution of complex organic molecules. The most important prebiotic condition for the beginning of biological evolution is the appearance of complex mixture of molecules with high diversity of structures. **Stuart Kauffman** has calculated that such critical diversity should comprise at least two hundred thousand molecules with different structures. It is only the element carbon which can form such enormous number of molecular structures. Today, more than ten million organic compounds are known.

In the second half of XIX century, **August Kekulé**, **Archibald Couper**, and (independently) **Aleksandr Butlerov** have discovered the most important property of carbon: interlinking of carbon atoms in chains, branched chains, and cyclic structures. This discovery was extended with the discovery that carbon atoms can be connected by stable single, double, and triple bonds. For comparison, silicon atoms can also be linked by single and double bonds, but in contrast to carbon–carbon bonds, silicon–silicon bonds are weaker, unstable, and sensitive to light. Hence, silicon-organic compounds would not survive under the environmental conditions on Earth during the period in which biological evolution begun. However, silicon atoms can form strong bonds with oxygen atoms. There is high diversity of structures in nature in which silicon–oxygen–silicon structural motifs are present. Such structures are characteristics of the mineral world, but they cannot be the base for the appearance of life.

$$-\overset{|}{\underset{|}{C}}-\overset{|}{\underset{|}{C}}-\overset{|}{\underset{|}{C}}-\overset{|}{\underset{|}{C}}-\overset{|}{\underset{|}{C}}-\overset{|}{\underset{|}{C}}-\overset{|}{\underset{|}{C}}-\overset{|}{\underset{|}{C}}-\overset{|}{\underset{|}{C}}-$$

$$-\overset{|}{\underset{|}{Si}}-O-\overset{|}{\underset{|}{Si}}-O-\overset{|}{\underset{|}{Si}}-O-\overset{|}{\underset{|}{Si}}-O-\overset{|}{\underset{|}{Si}}-$$

Later in this book, we will discuss the additional special properties which are responsible that carbon is a unique biogenic element.

Basic structural characteristics of organic molecules origin from the structures of the allotropic modifications of carbon. Until the last quarter of the XX century, only two allotropic modifications of carbon were known: **graphite** and **diamond**. The arrangements of atoms which appear in graphite and diamond represent the basic structural patterns by which carbon atoms can be interconnected.

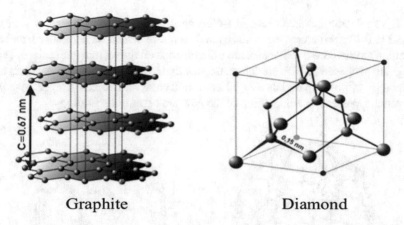

Graphite   Diamond

In graphite, every carbon is surrounded by three neighboring carbon atoms in such a way that all atoms lie in the same plane. In contrast, the carbon atoms in diamond are arranged in the three-dimensional array where every atom is surrounded by four neighbors which are configured in tetrahedral geometry. These two motifs, tetrahedral and planar trigonal, respectively, represent basic structural patterns of organic molecules. Linear binding of atoms is also possible for organic molecules, but allotropic modification of carbon with such structure has not been observed yet.

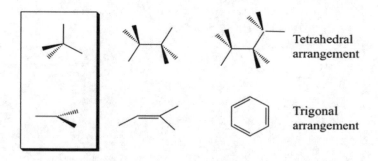

Tetrahedral arrangement

Trigonal arrangement

In the intergalactic space, there are stars which are in the last phase of their development and which produce a lot of elemental carbon by eruptions. **Harold Kroto** and his collaborators **Richard Smalley** and **Robert Curl** have investigated in detail the nature of such intergalactic carbon. The result of their research has been the discovery of new allotropic modification of carbon in which atoms form structures resembling a ball. By measuring the relative molecular masses of such ball molecules and simulating the interstellar conditions in the laboratory, Kroto, Smalley, and Curl have found that these molecules mostly consist of 60 carbon atoms interconnected in pentagonal and hexagonal structures. There are 12 pentagons surrounded by hexagons. Since the proposed structure resembles some works of art, especially the architecture invented by the architect **RichardBuckminster Fuller,** this C_{60} molecule has been named **fullerene**. In subsequent research, a series of similar ball-like structures were discovered, some of which have tubular structures of carbon atoms. These molecules have dimensions on the nanometer scale and have intriguing properties which are interesting for their applications in the sophisticated technology of novel materials as well as in molecular electronics. The discovery of fullerenes represents the beginning of the new era of **nanotechnology**.

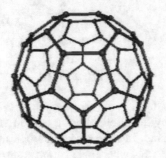

FULLEREN

Contents

Chapter 1
Alkanes, Composition, Constitution and Configuration

In principle, the organic molecules may be considered as consisting of hydrocarbon skeleton to which functional groups are attached. While the hydrocarbon skeleton is responsible for the molecular shape and flexibility, chemical reactivity depends mostly on the presence of functional groups. For better understanding basic properties and structure of molecular skeleton, we will start with the simple hydrocarbons which are the organic compounds without functional groups.

The compounds whose molecules consist of carbon and hydrogen, hydrocarbons can be divided in two main categories. In the first group are **saturated** hydrocarbons, **alkanes** and **cycloalkanes**, and in the second group are **unsaturated hydrocarbons**, alkenes, alkynes, as well as aromatic compounds. The term "saturated" means that the maximal possible number of hydrogens are present in the molecule. No additional hydrogens can be added. The following scheme shows how unsaturated molecule can be transformed in the saturated structure by addition of hydrogen. Ethene as the simplest unsaturated compound can be converted to the saturated ethane by addition of the hydrogen molecule. Since all the valences of carbon atoms in ethane are occupied with hydrogens, there is no place for binding any more hydrogen atom.

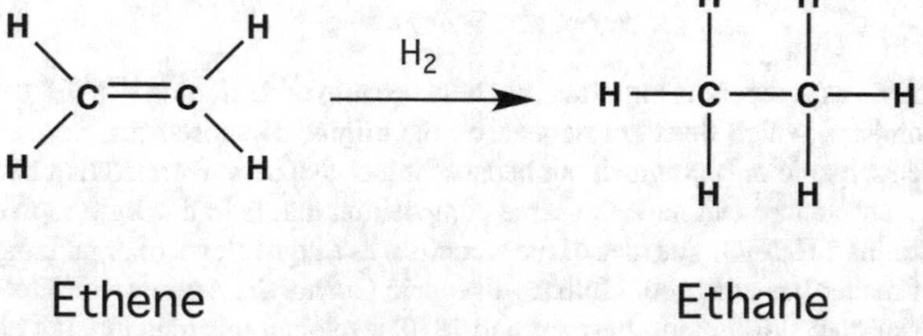

Ethene → (H₂) → Ethane

Nature is abundant with hydrocarbons, especially in crude oil and natural gas. The mixture of hydrocarbons present in oil can be separated into groups of compounds with different boiling points by using industrial procedures called fractional distillation. Analysis of compounds from different fractions provides their elemental **composition** from which chemical formulas can be calculated, for instance C_2H_6 for ethane. Determination of composition is based on the property of hydrocarbons to combust into water and carbon dioxide. At the beginning of XIX century, **Jöns Jacob Berzelius** has calculated formulas for a series of organic compounds starting with the measured masses of water and carbon dioxide.

Although the knowledge of composition is very important for the classification of organic compounds, for description of organic molecules in more detail it was nonetheless insufficient. Difficulties appeared because there are different organic compounds which have the same composition. Atomic theory that could help resolve these contradictions had still been in the early stages of development at the time. **John Dalton** has published his discovery of atoms only in 1808 in his book *The New System of Chemical Philosophy*.

There is additional controversy that has appeared during this period in the history of chemistry. It was believed that inorganic and organic compounds have different natural origins. Hence it was thought impossible that organic compounds could be prepared from any inorganic source. They, so it was thought, can be obtained exclusively from living organisms. This *vitalistic* point of view was disproved by the student of Berzelius, **Friedrich Wöhler** who has successfully prepared the organic compound from the inorganic precursor. The organic compound urea has been obtained by heating the inorganic salt ammonium cyanate:

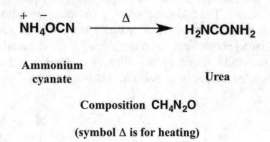

$$\overset{+}{N}H_4\overset{-}{O}CN \xrightarrow{\;\;\Delta\;\;} H_2NCONH_2$$

**Ammonium   Urea
cyanate**

Composition CH_4N_2O

(symbol Δ is for heating)

Wöhler's experiment is important not only because of its finding that there is only one chemistry, which does not depend on the origins of substances, i.e., whether they are inorganic or biological, but because it has also demonstrated that two very different substances can have the same composition that is in this case represented with formula CH_4N_2O. The idea of the **structure** as a higher level of organizing principle of matter has emerged. **Charles-Frédéric Gerhardt**, **August von Hofmann** and **Alexander Williamson** have around 1850 developed this idea into the concept of **constitution** which describes the way in which atoms are interconnected in the molecule. In Wöhler's experiment different *constitutions* of ammonium cyanate and urea can be described by different *structural formulas*:

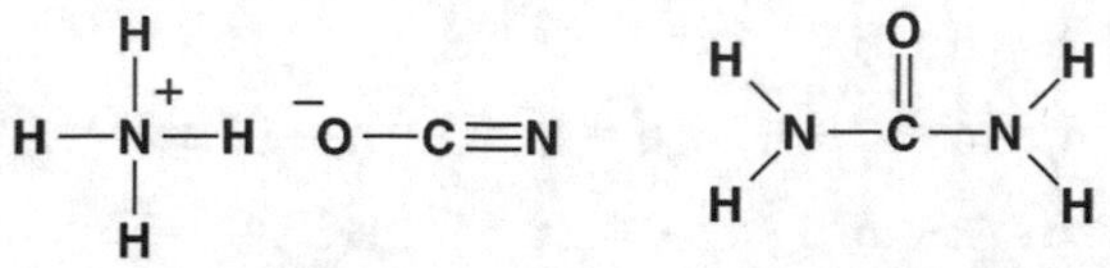

**Ammonium
cyanate**

Urea

Compounds such as urea and ammonium cyanate, which have the same composition but different constitution are called **isomers**. After this first example it has been found that in the nature exists enormous number of isomers of organic compounds especially hydrocarbons. Before clarifying these concepts in more detail let us describe composition, constitution of some of the most important saturated hydrocarbons, **alkanes**. As can be seen from the following Table, all the given chemical formulas can be reduced to the general formula C_nH_{2n+2}, where n is an integer. Such alkanes are called n-alkanes and belong to a **homologous series** of these compounds.

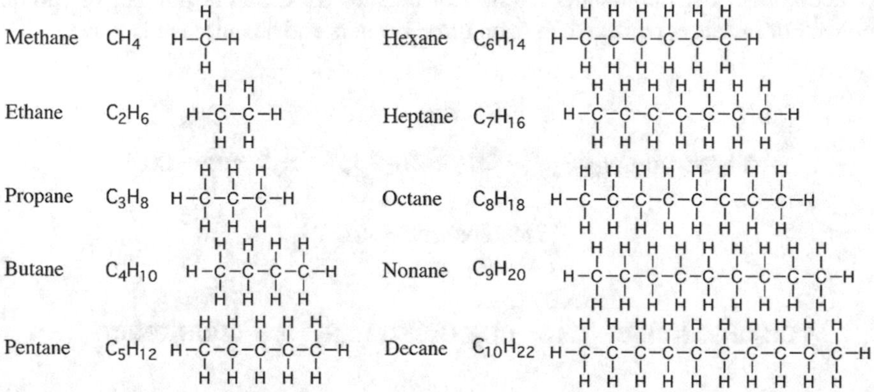

Constructions of different isomers are possible for the alkane molecules which contain more than three carbon atoms. In the molecule of butane with the composition defined by formula C_4H_{10}, the carbon atoms can be interconnected in two different ways forming two constitutional isomers. While the compound with the linear carbon chain is usually called n-**butane** its branched isomer is called iso-**butane**.

Structures shown in the following scheme are two different isomers which really represents two different compounds. To convert one isomer into another it is necessary to break and reform chemical bonds by chemical reaction. The isomers are represented in two ways, by structural formula in which all the interatomic bonds and atomic symbols are shown, or by using the condensed formula where only the chain of the CH_3 and CH_2 groups is drawn. Such more practical formulas are mostly used in organic chemistry. The particular groups in the chain have special names such as **methylene group** for CH_2 and **methyl group** for CH_3.

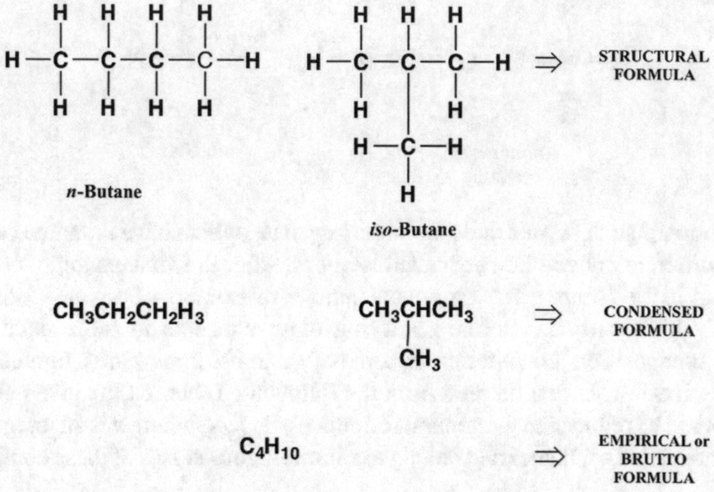

$$H-\underset{\underset{H}{|}}{\overset{\overset{H}{|}}{C}}-\underset{\underset{H}{|}}{\overset{\overset{H}{|}}{C}}-\underset{\underset{H}{|}}{\overset{\overset{H}{|}}{C}}-\underset{\underset{H}{|}}{\overset{\overset{H}{|}}{C}}-H$$

n-Butane

$$H-\underset{\underset{H}{|}}{\overset{\overset{H}{|}}{C}}-\underset{\underset{\underset{\underset{H}{|}}{C}-H}{\underset{H}{|}}}{\overset{\overset{H}{|}}{C}}-\underset{\underset{H}{|}}{\overset{\overset{H}{|}}{C}}-H \quad \Rightarrow \quad \text{STRUCTURAL FORMULA}$$

iso-Butane

$CH_3CH_2CH_2H_3$ $CH_3\underset{\underset{CH_3}{|}}{CH}CH_3 \quad \Rightarrow \quad$ **CONDENSED FORMULA**

$C_4H_{10} \quad \Rightarrow \quad$ **EMPIRICAL or BRUTTO FORMULA**

Let us consider the number isomers of alkanes for different number of carbon atoms. For instance, pentane can form three isomers and hexane five isomers:

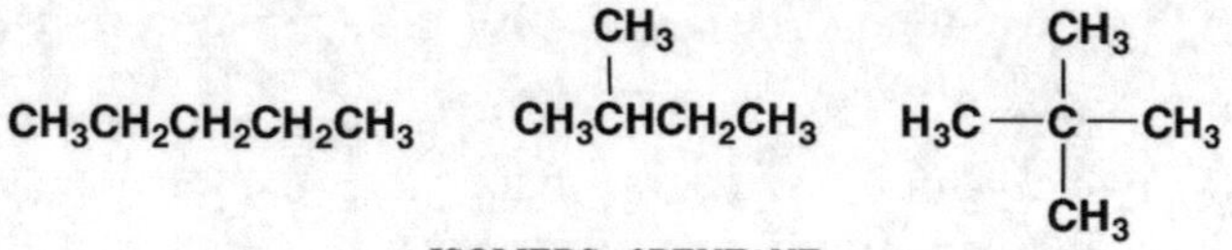

$$CH_3CH_2CH_2CH_2CH_3 \qquad CH_3\underset{\underset{CH_3}{|}}{CH}CH_2CH_3 \qquad H_3C-\underset{\underset{CH_3}{|}}{\overset{\overset{CH_3}{|}}{C}}-CH_3$$

ISOMERS of PENTANE

$$CH_3CH_2CH_2CH_2CH_2CH_3 \qquad CH_3CH_2\underset{\underset{CH_3}{|}}{CH}CH_2CH_3 \qquad CH_3\underset{\underset{CH_3}{|}}{\overset{\overset{CH_3}{|}}{CH}}CH_2CH_2CH_3$$

$$\underset{H_3C}{\overset{H_3C}{\diagdown}}CH-CH\underset{CH_3}{\overset{CH_3}{\diagup}} \qquad H_3C-\underset{\underset{CH_3}{|}}{\overset{\overset{CH_3}{|}}{C}}-CH_2CH_3$$

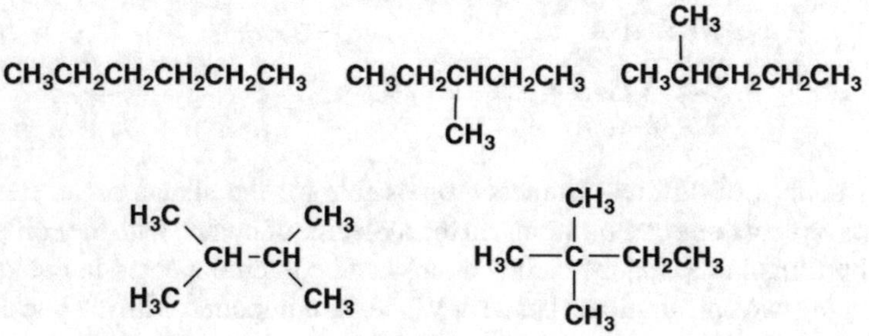

IZOMERS of HEXANE

In following table, the number of isomers is given for the alkanes with up to 10 C-atoms. The table shows that the number of isomers increases enormously with the number of carbon atoms in the alkane molecule. The saturated hydrocarbon with 20 carbon atoms with the brutto formula $C_{20}H_{42}$ can have 366,319 isomers!

Number of C-atoms in alkanes	Number of possible constitutional isomers
1	1
2	1
3	1
4	2
5	3
6	5
7	9
8	18
9	35
10	75

Such a large number of isomers of simple alkanes explains why carbon as an element is unique and why it serves as a basis for the vast diversity of structures necessary for the appearance of Life. This propensity for generating great diversity of organic molecules starting from simple structures will be exemplified further in this book.

Regarding isomers, it must be noted that depending on the way the atoms are interconnected, C-atoms can bind to each other in several different ways. Some carbons are bound to only one neighboring C-atom, some to two, three or four. Based on this criterion the carbon atoms are named **primary**, **secondary**, **tertiary** and **quaternary**, respectively:

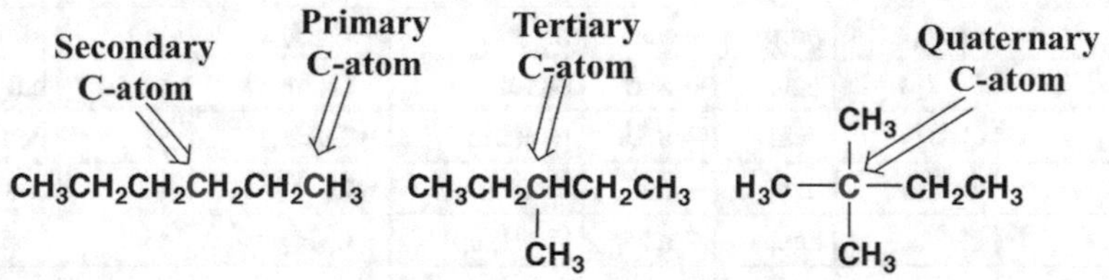

1.1 On the Nomenclature of Organic Compounds

When discussing chemistry as a discipline we must be aware of three different categories which are present in the methodology of chemical science and practice. Chemistry can be recognized by its **content**, **script** and **language**. The most important is the content; these are the substances with which we have immediate experience either in laboratory or in everyday life. Chemical script and language are human inventions; they are a kind of models which serve for more or less unambiguous communication between chemists about substances, chemical concepts and theories. While chemical script comprises formulas which we have already mentioned, by chemical language we describe constitutions and configurations of molecules.

The chemical language is designed to be sufficiently precise that from the name of a compound only one structural formula can be deduced. The names of compounds are based on the linguistic rules called **nomenclature**. Today, the chemical nomenclature is universal, standardized and governed by international conventions promulgated by the International Union for Pure and Applied Chemistry (IUPAC). According to IUPAC convention, the name of the compound consists from the root word to which the prefixes and suffixes can be added, depending on the class and structure of the molecule. The root word of the name is based on the number of C-atoms in the longest carbon chain and it is derived from the names of simple hydrocarbons.

The suffix is the label for the functional group characteristic for particular class of chemical compounds. In this scheme the saturated hydrocarbons, the **alkanes** have the suffix **–ane**. For naming isomers, the system is more complicated and includes additional rules. Since the molecules of isomers are branched the root word must correspond to the longest chain. The sidechains are treated as additional groups called **substituents**. In the final name of the structure, the substituents are introduced as prefixes to root word. The names of substituents are formed following the same rules as in the case of simple alkanes, i.e., the number of C-atoms with the added suffix **-yl**.

Number of C-atoms	Formula	Root	Suffix	Name of compound	Substituent	Suffix	Name
1	CH_4	met-	– ane	methane	$-CH_3$	-yl	methyl
2	C_2H_6	et-	– ane	ethane	$-C_2H_5$	-yl	ethyl
3	C_3H_8	prop-	– ane	propane	$-C_3H_7$	-yl	propyl
4	C_4H_{10}	but-	– ane	butane	$-C_4H_9$	-yl	butyl
5	C_5H_{12}	pent-	– ane	pentane	$-C_5H_{11}$	-yl	pentyl
6	C_6H_{14}	hex-	– ane	hexane	$-C_6H_{13}$	-yl	hexyl
7	C_7H_{16}	hept-	– ane	heptane	$-C_7H_{15}$	-yl	heptyl
8	C_8H_{18}	oct-	– ane	octane	$-C_8H_{17}$	-yl	octyl
9	C_9H_{20}	non-	– ane	nonane	$-C_9H_{19}$	-yl	nonyl
10	$C_{10}H_{22}$	dec-	– ane	decane	$-C_{10}H_{21}$	-yl	decyl

The position of the substituent on the longest chain is labeled by a number. Terminal carbon of the chain must be selected so that all other substituents have the smallest possible numbers. If on the same basic chain two or more identical substituents are attached, the suffix is expanded by adding labels di-, tri-, etc. For instance, in the name 2,2-dimethylpropane, (shown in the schemes below) the numbering 2,2- means that two methyl groups are bound to the carbon 2.

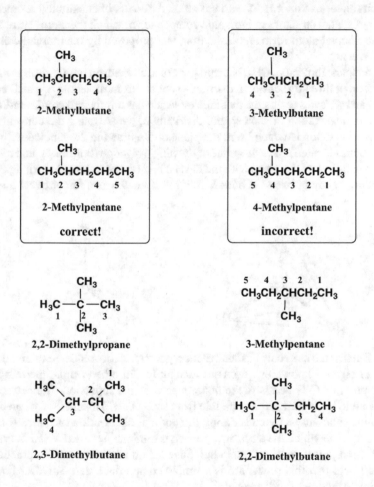

1.2 Configurations and Shapes of Molecules

The basic idea that molecules are real particles which have particular shape originated from three chemists, one of them was organic, the other two were inorganic and physical chemists. By studying the symmetry of crystals of the organic salt ammonium-sodium tartrate, which has been isolated from the reaction mixture in alcoholic fermentation, **Charles LeBel** who worked with **Louis Pasteur** proposed in 1874 that the atoms bound to the central carbon atom in substituted alkanes are distributed in space so as to form tetrahedron. Such tetrahedral spatial configuration resembles the distribution of C-atoms in diamond. Details of this LeBel's discovery will be discussed later in this book. The same idea about the tetrahedral structure of the alkane-like molecules has been independently proposed by physical chemist **Henricus Jacobus van't Hoff** who has studied isomers of substituted alkanes. The concept of spatial structure of inorganic compounds in which the atoms surrounding the central metal atom form an octahedron, was proposed by the inorganic chemist **Alfred Werner**.

The methane molecule, CH_4, has the shape of a tetrahedron with carbon atom in its center. Spatial three-dimensional distribution of atoms in molecule is called **config-uration** and we can say that the methane molecule has tetrahedral configuration. For pictorial representation of such spatial distribution, we are using the convention by which chemical bonds which lie in the plane of the drawing are labeled with a full line, the bonds located above the plane of drawing by wedge (bold elongated triangle) and the bonds below the plane with the dashed line (dashed elongated triangle). The angle between any of two C-H bonds, $109°28'$, is known as a **tetrahedral angle**.

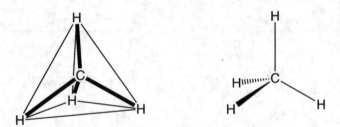

The fact that the molecule has such distinct geometric form can be explained by the branch of physics known as **quantum mechanics**. In other words, the tetrahedral configuration of C-H bonds is the consequence of the repulsion of electron pairs which tend to be as far apart as possible from each other. This method of prediction of the molecular shape by considering the optimal distribution of bonds (bonding electron pairs) in which the electron repulsion is minimal, is called **VSEPR** (**v**alence **s**hell **e**lectron **p**air **r**epulsion). Although this method is widely used in practice, we must point out that this procedure is a simplified approach that can afford only an approximate picture of the molecule.

Let us start with this knowledge about the tetrahedral shape of molecules to build structures of other simple alkane molecules. We can notice that the consequence of tetrahedral structure is the zig-zag form of the alkane chains. Chemical formulas in the following scheme are called wedge-dash formulas.

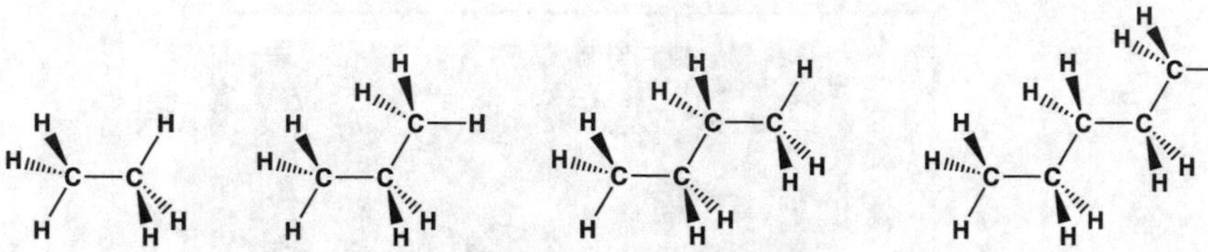

Bearing this shape of in mind it is possible to write the structural formula in an even simpler form. The figure below shows the structures of alkanes, their simplified structural formulas as well as their condensed structural formulas.

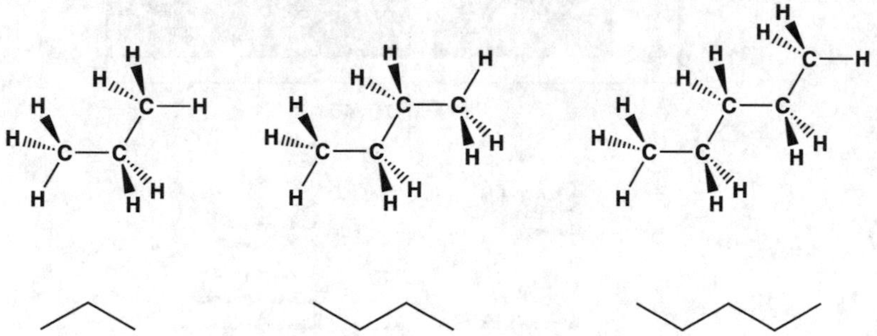

$CH_3CH_2CH_3$ $CH_3CH_2CH_2CH_3$ $CH_3CH_2CH_2CH_2CH_3$

Although simplified, this geometry model explains molecular shapes enough satisfactorily to represent an approximate picture of molecules. More detailed insights into the molecular shapes is possible by using special microscopy techniques called **scanning tunneling microscopy (STM)** or by complicated and sophisticated quantum mechanical calculations.

It is interesting to mention that there is a correlation between molecular structures of alkanes and some of their physical properties. By correlating the number of carbon atoms in simple alkanes with the melting points of the same compounds we observe that the molecules with odd numbers of C-atoms and those with even numbers of C-atoms exhibit different correlation curves.

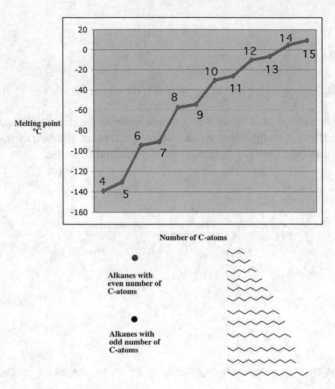

Looking at the structures of alkanes in the figure we can see that while the molecules with an odd number of carbon atoms appear more symmetrical (both terminal carbon–carbon bonds being oriented upwards-black lines), but the structures with an even number have the terminal carbon–carbon bonds oriented upwards on the left end of the chain, but downwards on the right end (blue colored formulas). Although the correct explanation is not simple, this example demonstrates how macroscopic properties correlate with the representations of microscopic structures. We can assume that "odd " molecules in the condensed state are packed differently from "even " molecules.

1.3   Molecular Dynamics and Conformations

Consideration of long-chained alkane molecules leads to the question of whether these molecules are flexible? For analyzing this flexibility, let us use the ethane molecule as an example. Starting with tetrahedral configuration, the ethane molecule could be represented in two different ways:

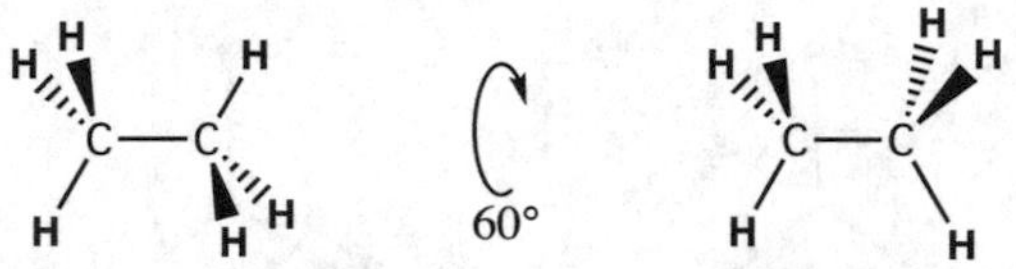

Form A		**Form B**
Staggered		**Eclipsed**
conformation		**conformation**

The form A can be transformed into B simply by rotation of one of the methyl groups by the angle $\Phi = 60°$ around the C–C bond. This angle of 60° is called the **torsion angle**. Such rotations, especially at room temperature, are very fast and it is possible to imagine that the ethane molecule could appear in a large number of shapes, depending on the rotation angle Φ. Such different shapes of molecules which follow from internal rotations about single bonds are called **conformations**. In the figure above the two most important conformations: **staggered** and **eclipsed** are shown using wedge-dash notation. However, these conformations can also be drawn by rotating the molecule by 90° relative to the plane of drawing. In that case the carbon atoms appear one behind the other. As it is shown in the following figure, the carbon atoms in the front and at the back are separated by a circle. Such representation is called **Newman formula**.

Newman formulas serve to clarify the difference between staggered and eclipsed conformations. In the staggered conformation, the C-H bonds at the neighboring carbon atoms are closer to each other than in the eclipsed conformation. Since the electron clouds in these covalent bonds are negatively charged, they lead to repulsive interactions so that the eclipsed conformation has higher potential energy than the staggered conformation. We can say that the neighboring C-H bonds **sterically hinder** each other, which gives molecule in the eclipsed conformation higher potential energy.

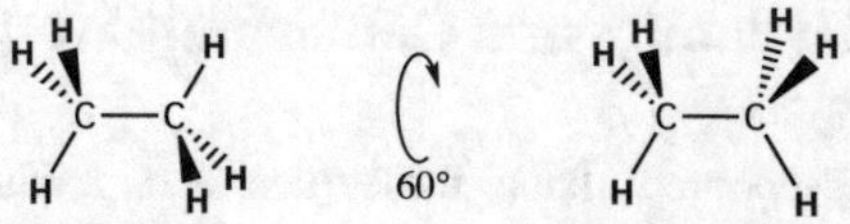

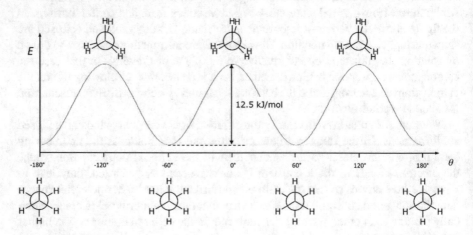

Form A
Staggered
conformation

Form B
Eclipsed
conformation

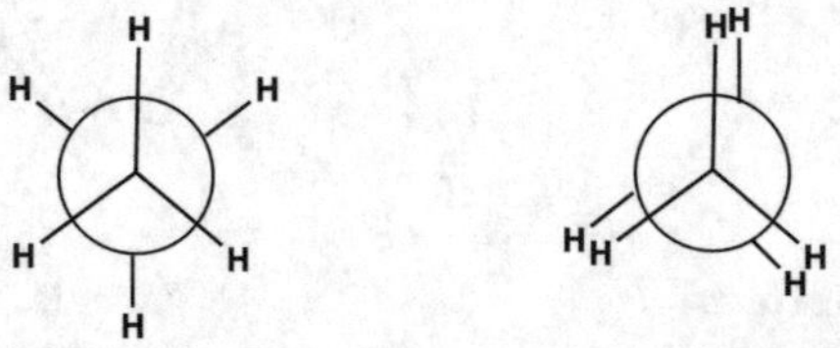

NEWMAN FORMULA

Starting with the eclipsed conformation, if the torsion angle increases the potential energy decreases up to the angle of 60° which is characteristic for the staggered conformation. Such dependence of the potential energy on torsional angle is shown in the following scheme.

We can observe that the estimated potential energy varies periodically with the angle and that staggered conformations always correspond to the energy minima while eclipsed conformations always corresponding to the energy maxima. It is for this reason that ethane molecules adopt the staggered rather than the eclipsed conformation.

The structures which correspond to the energy minima are called **conformers**. If the molecule comprises more than one C–C bond the rotation is possible around all of them and the number of conformers increases. Let us investigate the conformations and conformers of the butane molecule:

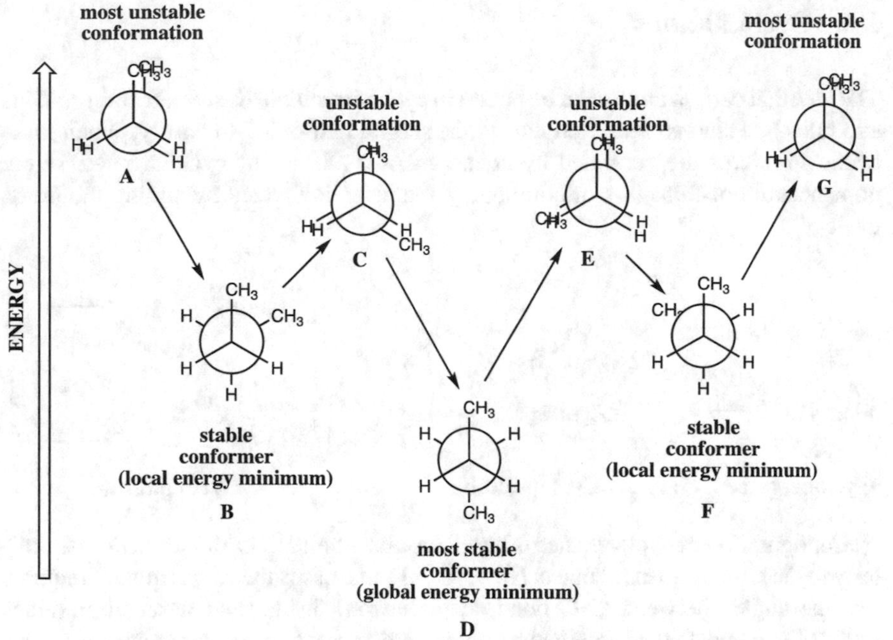

By investigating the conformations of butane and considering only the rotation about the central C–C bond we can see that three conformers are possible (labeled **B**, **D** and **F** in the scheme above). The most stable conformer is **D** because in this case the bulky methyl groups which have dense electron clouds are at the largest distance from each other, which makes their **steric hindrance** smallest. Conformer **D** has the smallest potential energy and we can say that it corresponds to the **global minimum** of the potential energy curve. Less stable conformers **B** and **F** correspond to **local minima** of the potential energy curve. Transformation of one conformer into another is possible only by passing through the potential energy maxima **A**, **C**, **E** and **G**, which belong to eclipsed conformations. As the molecule gets more complex, the number of conformers (local energy minima), as well as the number of local maxima (eclipsed conformations) becomes larger. Exploring the shapes and energies of possible conformers and the energies of eclipsed conformations requires special calculations called **conformational analysis**. The simplest and most often used method is based on the mechanical model in which the atoms are defined as mass points and the chemical bonds are treated as elastic connectors. Using this model, it is possible to calculate energy minima and maxima. The method is called **molecular mechanics**.

1.4 Cycloalkanes

The idea that carbon atoms can be bound in cyclic structures (besides forming chains and branched chains) appeared during the second half of XIX century. When these cyclic structures are saturated hydrocarbons, we talk about **cycloalkanes**. In the nomenclature of this class of compounds the name is formed by adding the prefix **cyclo-**.

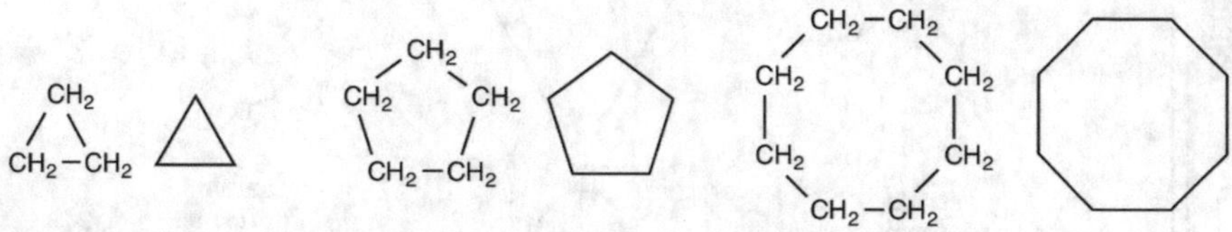

Cyclopropane   **Cyclopentane**   **Cyclooctane**

Configuration of cycloalkanes is based on the combinations of tetrahedrons, similarly to the case of linear alkanes. However, in some cases the ring structure requires that the angles between C–C bonds deviate from the normal tetrahedral values (109°28′). Almost ideal tetrahedral angles are present in the molecule of cyclohexane. For the pictorial representation of three-dimensional molecular structures of cyclic molecules special descriptive projection is used. Let us analyze the structure of cyclohexane molecule in more detail.

1.4.1 Cyclohexane

As we can see in the following scheme the hydrogen atoms can be bound to carbons in two different ways. When C-H bonds (containing green H atoms) are parallel to the axis perpendicular to the plane of the ring, hydrogens in such bonds are called **axial** hydrogens (shown in green). Hydrogens drawn in red lie nearly in the equatorial plane of the ring and are called **equatorial** hydrogen atoms. The difference between these types of hydrogens is better represented by using also the Newman formula:

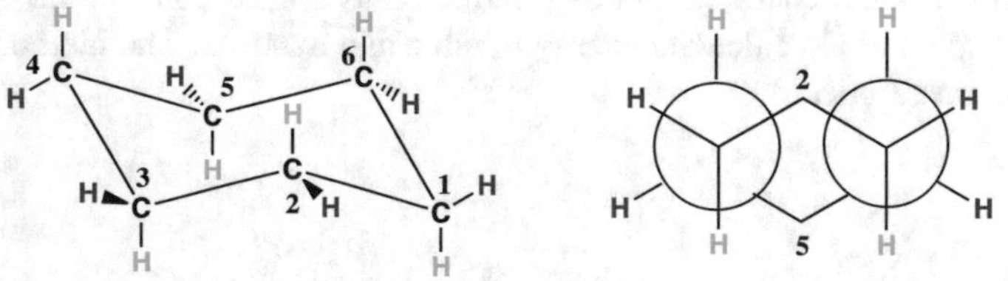

The Newman formula shown in the upper scheme (right) is obtained by looking along the C_3–C_4 and C_1–C_6 bonds. On the deeper inspection it becomes clear that the represented conformation has staggered form and therefore belongs to minima on the potential energy surface. By convention such conformers are called **chair conformers**. Since all the C–C bonds are in the staggered arrangements the cyclohexane molecule is the most stable cyclic hydrocarbon. There are no C–H bonds that sterically hinder each other. Conformational dynamics of the cyclohexane ring involves limited rotation around the C–C bonds: one chair conformer is transformed into another.

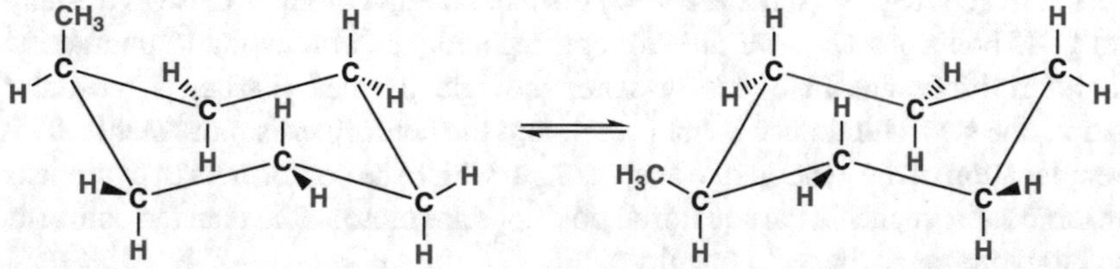

Hydrogen atoms (blue) that are in the axial positions in one conformer appear in equatorial positions in the product conformer. Such change from axial to the equatorial position is even more evident in the methyl-substituted cyclohexane.

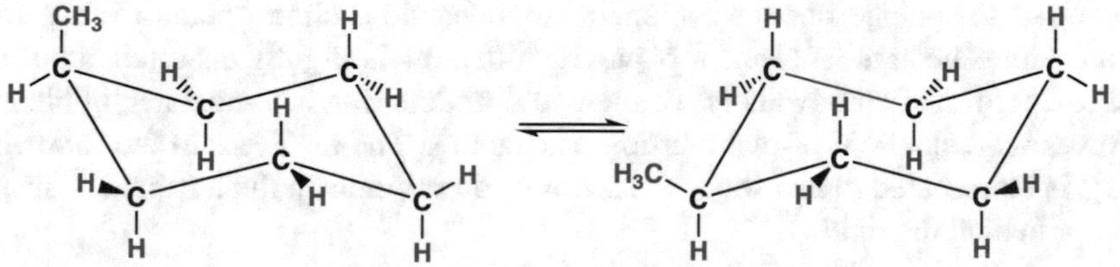

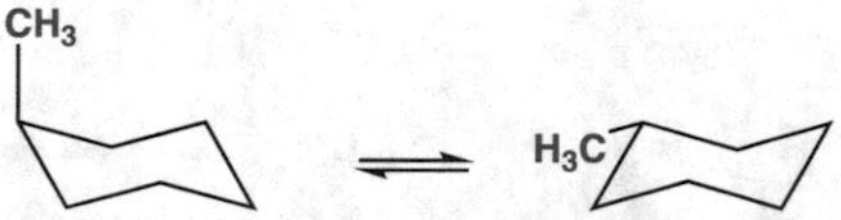

More stable chair conformation

The methyl group is in the axial position in the first conformer and in equatorial position in the second conformer. Both conformers can also be represented by Newman formulas:

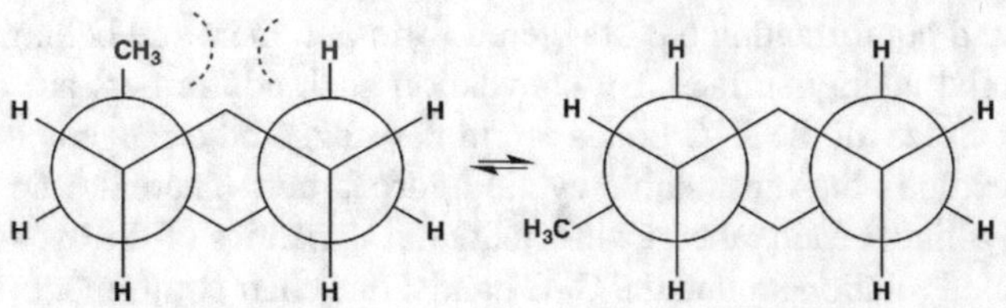

More stable conformer

In the left conformer (scheme above) the axial methyl group is close to the nearby axial C–H bond. As we have already commented, such unfavorable interaction is called steric hindrance. In the conformer on the right, the methyl group is in equatorial position, the steric hindrance is not preset, and the conformer is more stable. This is the explanation of the general rule according to which the conformers of cyclohexane with large substituents in the equatorial positions are more stable than the conformers with large groups in the axial positions.

1.4.2   Cyclopentane

In contrast to cyclohexane the cyclopentane molecule is almost planar. While four carbon atoms lie in the molecular plane the fifth atom is slightly distorted out of the plane. Conformational dynamics of the cyclopentane ring is almost negligible and involves sequential out-of-plane lifting of C-atoms. The molecule in the following figure is represented in two ways, the accurate description on the left-hand side and a simplified on the right.

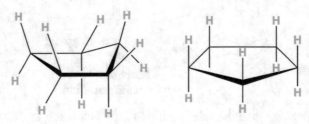

Simplified representation

Since, in average, all the hydrogen atoms are quasi-axial the representation of cyclopentane molecule by planar ring is an acceptable description.

1.4.3 Cyclobutane and Cyclopropane

While the angles around carbon atoms in cyclohexane and cyclopentane do not deviate much from the tetrahedral angle, in the small ring molecules such as cyclobutane or cyclopropane the geometry requires that the angles between neighboring C–C bonds deviate significantly from the tetrahedral value. Such forced reduction of tetrahedral angles requires additional potential energy. Consequently, small cyclic molecules such as cyclopropane contain an excess of potential energy that is called the **strain energy** or the **Bayer strain**. Because of the strain, the molecules with three and four membered rings have an excess of energy and they are chemically active. In chemical reactions the ring tends to cleave and relieve excess of the potential energy.

Cyclobutane   **Cyclopropane**

The cyclobutane molecule is not planar and carbon atoms deviate from the plane by forming the structure in which four of the hydrogen atoms have equatorial position (labeled in red) and the remaining four hydrogens are axial (labeled in green). In this way the Bayer strain is minimized. In cyclopropane, all the hydrogens are in quasi-axial positions.

From the detailed studies of cyclopropane molecule, it has been found that the C–C bonds are folded i.e., the electron density maxima between bonded carbon atoms lie off the C–C bond axis.

1.5 Polycyclic Hydrocarbons

Carbon atoms can also be interconnected in structures that have more than one ring. They are called **polycyclic hydrocarbons**. One well known polycyclic structure that appears in a series of natural products is **decalin** with two condensed six membered rings. The term «condensed» means that rings share one C–C bond. Since there are two different orientations of hydrogens on this common C–C bond, the decalin molecule can have two stereoisomers. The neighboring hydrogens can be either on the same side or on the opposite sides of the shared C–C bond and the stereoisomers are known as *cis*-**decalin** and *trans*-**decalin**, respectively. As expected, the isomers have different physical properties. The melting point of *cis*-decalin is 198 °C and the melting point of *trans*-decalin is 185 °C.

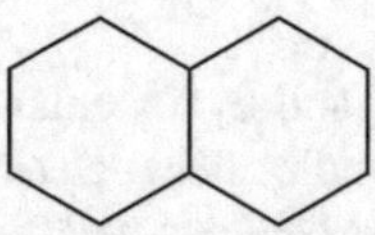

Decaline 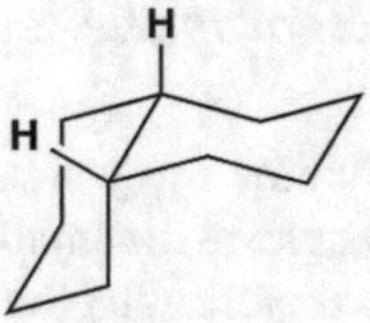   ***cis*-Decaline** 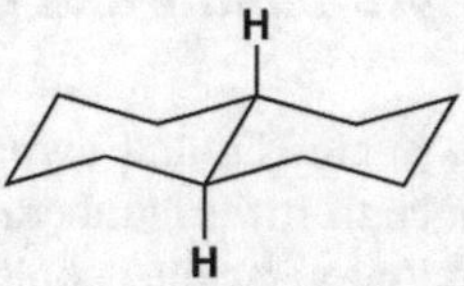   ***trans*-Decaline**

In chemical nomenclature, the polycyclic hydrocarbons are named regarding the size of the bridges. The carbon atoms which belong to two different rings are called **bridgeheads** and they are not considered as part of the bridge. The suffix **bicyclo-** is used if there are two rings and the number of C-atoms on the bridges is added in parenthesis.

Bridgehead carbon atom

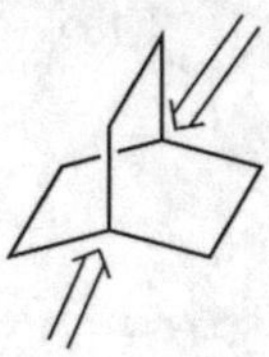

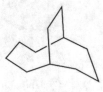

Bridgehead carbon atom

Bicyclo[2.2.2]octane   **Bicyclo[4.3.2]undecane**

Let us describe some frequently encountered polycyclic structures. **Bicyclo[2.2.1]heptane** also known as **norbornane** is the basic structure of a series of natural products. Tricyclic compound **adamantane** is in principle the smallest structural fragment of diamond and its structure appears on the list of pharmaceutically important products. Adamantane has been synthesized for the first time in 1941 at the University of Zagreb by Nobel Prize winner **Vladimir Prelog**.

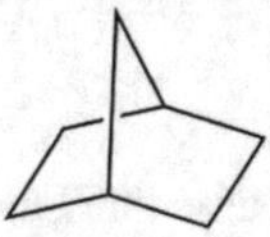

Bicyclo[2.2.1]heptane
(Norbornane)

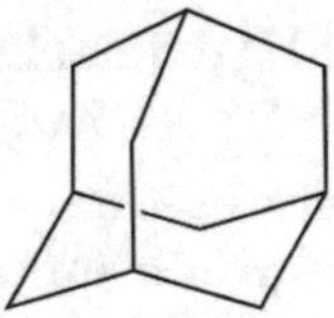

Adamantane

Chapter 2
Functional Groups

As we have seen in the previous chapter, the hydrocarbon skeleton is responsible for shape and flexibility of organic molecules. In the case of alkane molecules, the molecular structure is based on tetrahedral units and the molecular dynamics is the consequence of relatively free rotations about the carbon–carbon single bonds. These rotations give rise to different conformations. However, with the exception of small-ring molecules, the alkanes as the compounds containing only carbon and hydrogen are relatively weakly reactive substances.

Most of organic molecules which exhibit chemical reactivity have the active structural unit called **functional group**. In the structural formula the unspecified group, **the substituent**, bound to the hydrocarbon skeleton is labeled as R. To be chemically active, the functional group must possess high energy electrons which can be either the electrons in the multiple bonds or the non-bonded electrons on atoms other than carbon or hydrogen. Such atoms (for instance O, N, S, P, Cl, Br, I etc.) when present in the organic molecules are called **heteroatoms**. The presence of the functional group is also the basis for systematization of organic compounds into specific classes.

The most common functional groups together with their nomenclature are listed in following tables. The additional functional groups as well as the details of the nomenclature of specific classes of compounds will be discussed later in this book.

Amongst the functional groups which have double and triple bonds we shall mention those in the following table. All the compounds belong to different types of hydrocarbons:

© The Author(s), under exclusive license to Springer Nature Switzerland AG 2022   19
H. Vančik, *Basic Organic Chemistry for the Life Sciences*,
https://doi.org/10.1007/978-3-030-92438-6_2

ALKENES	-ENE	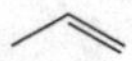	PROPENE
ALKYNES	-INE		PROPYNE
AROMATIC COMPOUNDS			BENZENE

The main functional groups and the corresponding organic compounds containing oxygen are alcohols, ethers, ketones, aldehydes, carboxylic acids and esters. They are listed in the following table:

ALCOHOLS	-OL	CH_3CH_2OH	ETHANOL
ETHERS	ETHER	$CH_3CH_2OCH_2CH_3$	DIEHYL ETHER
ALDEHYDES	-AL	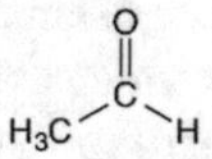	ETHANAL
KETONES	-ONE	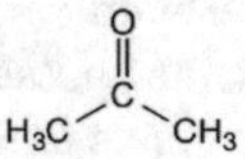	PROPANONE
CARBOXYLIC ACIDS	-OIC ACID	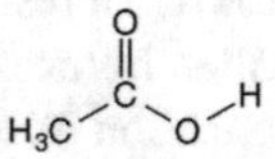	ETHANOIC ACID
ESTERS	-YL -OATE	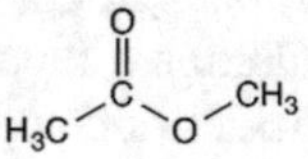	METHYL ETHANOATE

The most important organic compounds with nitrogen are the following:

AMINES	-AMINE AMINO-	$CH_3CH_2NH_2$	ETHYL AMINE AMINOETHANE
NITRILES	-NITRILE	CH_3CN	(ETHANONITRILE) ACETONITRILE
AZO COMPOUNDS	AZO-	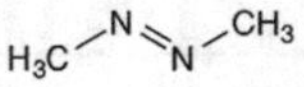 $H_3C-\underset{N}{\overset{N}{\diagdown}}\diagdown CH_3$	AZOMETHANE
DIAZO COMPOUNDS	DIAZO-	$H_2C=N^+=N^-$	DIAZOMETHANE

Some organic compounds have groups with both elements, oxygen and nitrogen:

AMIDES	-AMIDE	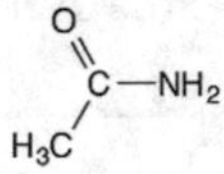 $\underset{H_3C}{\overset{O}{\underset{\|\|}{C}}}-NH_2$	ETHANAMIDE
NITRO COMPOUNDS	NITRO-	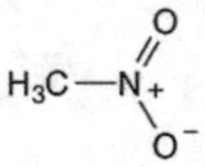 $H_3C-\overset{O}{\underset{O^-}{\overset{\|\|}{N^+}}}$	NITROMETHANE
NITROSO COMPOUNDS	NITROSO-	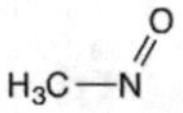 $H_3C-\overset{O}{\overset{\|\|}{N}}$	NITROSOMETHANE

Let us also list the classes of compounds with the functional groups containing sulfur and halogen:

HALIDES	-IDE HALO-	CH_3Cl	METYL CHLORIDE CHLOROMETHANE
THIOLS	-THIOL	CH_3CH_2SH	ETHANTHIOL
SULFIDES	-SULFIDE	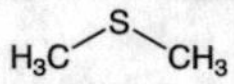	DIMETHYL SULFIDE
SULFOXIDES	-SULFOXIDE	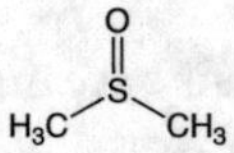	DIMETHYL SULFOXIDE
SULFONES	-SULFONE	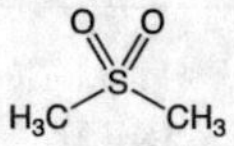	DIMEHYL SULFONE

Chapter 3
Electronic Structure of Organic Molecules

3.1 The Covalent Bond

More than a hundred years ago, **Joseph John Thomson** who has in 1897 discovered the electron, hypothesized that this small negatively charged particle plays crucial role in the formation of chemical bond. Thomson argued that atoms in the chemical bond exchange electrons in such a way that every atom donates one electron to the common chemical bond. Consequently, the electron pair is responsible for holding the two atoms together. However, the proposed model was not able to explain why the bonds in molecules containing the same atoms such as H_2 are different from the bonds in molecules with different atoms, for example HCl or NaCl. The idea about covalent and ionic bonds appeared in 1916 when **Gilbert Newton Lewis** proposed the representation of chemical bond as the common electron pair shared by the two bound atoms. The electrons in Lewis pairs do not have any identity regarding the atoms from which they originate and the covalent bond can be represented by dots as follows:

$$\textbf{H} \cdot \qquad \cdot \textbf{H} \qquad \textbf{H} : \textbf{H}$$

Using this model, the methane molecule is represented as:

$$
\begin{array}{c}
\textbf{H} \\
\cdot\cdot \\
\textbf{H} : \textbf{C} : \textbf{H} \\
\cdot\cdot \\
\textbf{H}
\end{array}
$$

Traditionally used valence lines acquire new meaning: they represent electron pairs. Within this semantics the double bond is labeled by two lines which describe two electron pairs. Commonly, such notation is called **Lewis structure**.

© The Author(s), under exclusive license to Springer Nature Switzerland AG 2022   23
H. Vančik, *Basic Organic Chemistry for the Life Sciences*,
https://doi.org/10.1007/978-3-030-92438-6_3

$$H_2C=CH_2 \qquad \cdot\cdot C \cdot\cdot C \cdot\cdot$$

However, in some molecules such as ammonia, some of the electrons are not included in the chemical bonds. Out of five valence electrons on the nitrogen atom, only three take part in covalent bonds with hydrogens, the remaining two are **nonbonding** and are called the **lone electron pairs**.

$$H—\overset{\cdot\cdot}{N}—H \atop | \atop H$$

Molecules with electron lone pairs are very reactive. Since organic chemistry is based on carbon compounds and the carbon atoms have electron lone pairs only in special cases, the carriers of nonbonding electrons are heteroatoms such as N, O, S, P or halogens.

$$CH_3CH_2\overset{\cdot\cdot}{\underset{\cdot\cdot}{O}}H \qquad CH_3CH_2CH_2\overset{\cdot\cdot}{N}H_2 \qquad CH_3—\overset{\overset{\displaystyle CH_3}{|}}{\underset{\underset{\displaystyle CH_3}{|}}{C}}—\overset{\cdot\cdot}{\underset{\cdot\cdot}{Cl}}:$$

The examples of carbon atoms with electron lone pairs are cyanide ion:CN^- or the relatively unstable but reactive molecules **carbenes**, which are common intermediates in photochemical reactions.

Although the Lewis model has been accepted as basic and universal concept for the description of constitutions of molecules, this representation appears to be inadequate for some species. For instance, by using Lewis model for the description of nitrate anion NO_3^-, as shown in the following figure, one of the oxygen atoms is bound to the nitrogen via double bond and two remaining oxygens via single bonds. Since it is known that double bonds are shorter than single bonds (this will be discussed later in the book) the proposed description of nitrate ion indicates that one of the nitrogen–oxygen bonds should be shorter than the two remaining bonds. However, since the precise measurements show that all three nitrogen–oxygen bonds are of equal length, the constitution of the nitrate ion cannot be explained by a single Lewis formula. At the beginning of the twentieth century **Sir Robert Robinson** and **Fritz Arndt** have proposed that such molecules can be represented with a set of structural formulas, which have different electron configurations. The particular electron configuration is called the **resonance structure** and we could say that the molecule is better represented with several resonance structures. In literature we can find the statement that the molecule is a **resonance hybrid** of the corresponding canonic resonance structures. In our example, the nitrate ion is better represented by three canonic resonance structures. By definition, all resonance structures together represent the same molecule. We use the notation in which different resonance structures are connected by double tipped arrows with all the formulas placed within the parentheses.

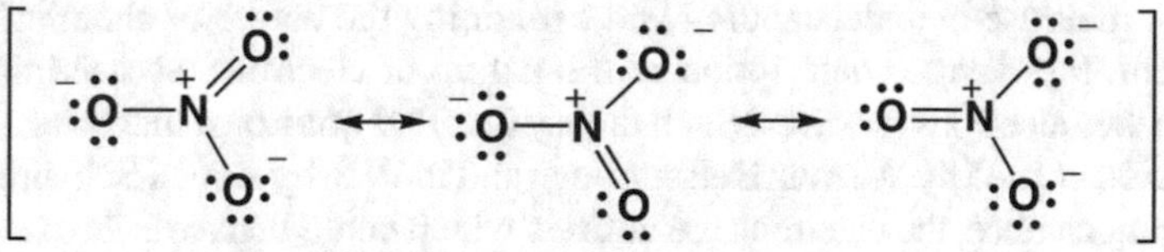

As shall be demonstrated later in this book, the knowledge of resonance structures can help in predictions of important molecular properties such as charge distribution or the nature of particular bond. Canonic resonance formulas can be constructed by using special rules:

1. Since the canonic structures belong to the descriptions of the same molecule, the atom positions in all the resonance formulas must be unchanged.
2. The total charge must also be unchanged. For instance, in the nitrate ion the total charge in all the structures is -1.
3. Only the electrons and electron pairs can be shifted. In principle, the electron lone pair in one resonance structure becomes the bonding pair in another structure and *vice versa*:

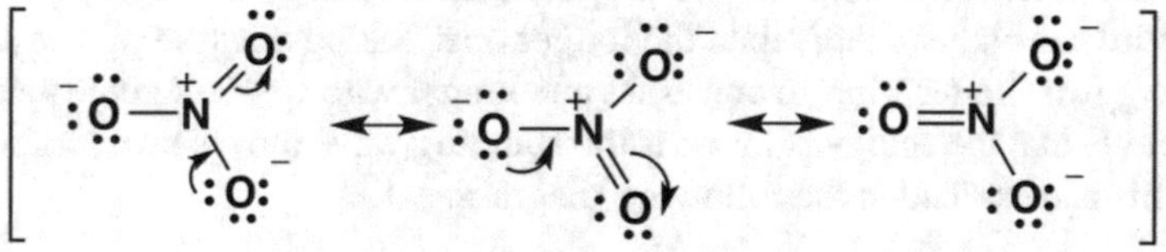

Robinson arrow

By convention, as a help in the representation of resonance structures, the shift of electron pairs is indicated by special arrows called **Robinson arrows** in the honor of **Robert Robinson**, the chemist who developed the theory of chemical reaction mechanisms.

There is an additional conclusion which can be drawn from the resonance model. The canonical resonance forms suggest that electron pairs do not have fixed positions within the molecule; they appear either as lone pairs or as double bonds. This demonstrates that electrons are distributed throughout the entire molecule or at least throughout its major part; the electrons are **delocalized**. It is important to point out that the **delocalization** of electrons decreases their energy as well as the total energy of the molecule. Molecule with delocalized electrons has lower energy than the molecule in which electrons are localized (for instance in single bonds or as lone pairs on the heteroatom). In this way, by constructing the canonic resonance formulas we are able to compare the stability of different molecules.

3.2   Molecular Orbitals

Although the Lewis concept together with the resonance model forms the basis for unambiguous descriptions of structures of molecular classes and for organic reactions, their use in understanding and explaining the nature of chemical bond is insufficient. For detailed description of the nature of electrons we need the theory that describes electrons as waves. Such theory is called **quantum mechanics** and has been created in 1925 by **Werner Heissenberg** and in 1926 by **Erwin Schrödinger**. In quantum mechanics, the electrons are entities which could behave either as particles or as waves, depending on the type of the experiment by which we observe them. We say that the electrons have **dual nature**.

For studying the electrons in atoms and molecules it is more convenient to consider electrons as waves. Since the electrons are positioned within atoms or molecules, the waves that describe them are **standing waves**, as are for instance the waves of water which move within the closed pool. Standing wave also describes the vibration of the strained wire, for example, on some musical instrument. For the pedagogical reason, let us use the strained wire vibrations as an analogy of electrons as waves. This could help in explaining the basic principles of the behavior of electrons in molecules, especially the role of symmetry.

After triggering, the wire vibrates by certain frequency and we hear the sound of the corresponding tune. If the wire is shorter the tune (frequency) is higher and vice versa, the longer the wire the lower the tune (frequency). As we know from physics, the vibrational frequency is proportional to energy and consequently, the shorter wire has higher energy than the longer one. Taking this rule into account, by connecting two shorter wires to construct one longer wire we shall get lower energy. We conclude that the energy is lower if the standing wave moves over larger area of space. This is described in the following diagram:

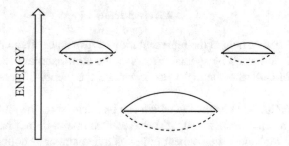

If the wave function is labeled with Φ, the lower energy wave is obtained by adding the functions Φ_1 and Φ_2. Imagine that the wires described by Φ_1 and Φ_2 are unified into the single wire which has twice the length and that we press the resulting longer wire in the middle (as we do for instance when playing the string instrument). The resulting wire will then vibrate with the opposite phases left and right from the pressure point. Such system could be represented as the difference of the functions

Φ_1 and Φ_2. The result is two composite waves different in phase but close to each other in energy. Let us represent them by $\Phi_1-\Phi_2$. If the smaller wire-waves are both electrically charged with the same charge (for instance negative), their common energy should be higher since the negative charges repel each other.

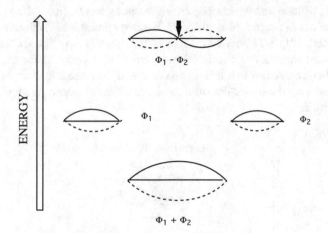

In principle, the two resulting standing waves can be obtained from Φ wave functions in two ways: in phase combination ($\Phi_1 + \Phi_2$) with lower energy and out of phase combination ($\Phi_1-\Phi_2$) with higher energy. Now, let us move to real systems where in atoms the standing waves, which describe electrons are represented by special functions called **atomic orbitals**. Here we will not discuss details of such mathematical functions. Rather we will use their graphic representations. Since carbon and hydrogen are the most important atoms in organic chemistry, we will represent their atomic orbitals only. While the electron in the hydrogen atom is present in only one orbital (1s), the electrons in the carbon atom are distributed in **1s, 2s, 2p$_x$, 2p$_y$ and 2p$_z$ orbitals**. From the following figure is clear that 2p$_x$, 2p$_y$ and 2p$_z$ orbitals are of equal energy which is however higher than the 2s orbital energy.

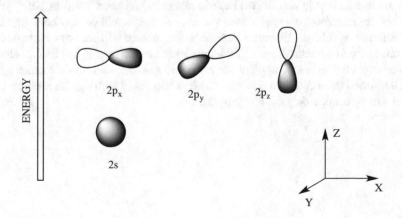

The p orbitals ($2p_x$, $2p_y$ and $2p_z$) differ only by their orientation in space along the coordinate axes x, y and z. Different color of the two sides of each p-orbital (gray or white) corresponds to different wave phases, similarly as it has been represented in the wire model analogy.

Next, we have to combine atomic orbitals to construct the molecule in the same way as we have done with wires representing standing waves. In the simplest case, the formation of the hydrogen molecule, we have two combinations of atomic orbitals: the lower energy "1s + 1s" combination and the higher energy "1s−1s" ombination. These two combinations of atomic orbitals describe electrons in the molecule and they are called **molecular orbitals**. In analogy with our standing wave model, these combinations i.e., the molecular orbitals, have different phase relations, **bonding** and **antibonding**:

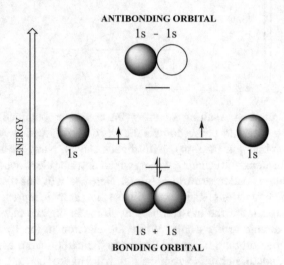

The appearance of electrons in the antibonding orbital disconnect covalent bond between two atoms. This can explain why two helium atoms cannot form covalent bond and the molecule He$_2$. If two helium atoms would form covalent bond, since each helium atom has two 1s electrons the four electrons will occupy both, bonding and antibonding orbitals. Electrons in antibonding orbital will destroy covalent bond converting the hypothetical helium molecule back into corresponding helium atoms. Similar explanation can be applied for all noble gas elements. On the other hand, each hydrogen atom contributes to the covalent bond only with one electron, and the two electrons occupy only the bonding orbital.

ENERGY

"1s - 1s"   "1s ↿ 1s"

1s ⥮   ↿ 1s   1s ⥮   ⥮ 1s

"1s + 1s"   "1s + 1s"

H₂   He₂

Bonding between carbon atoms could be described analogously with the exception that $2p_x$ orbitals are used instead of 1 s orbitals. Combining two $2p_x$ orbitals we can construct the bonding ("$2p_x + 2p_x$") and antibonding ("$2p_x - 2p_x$") molecular orbitals:

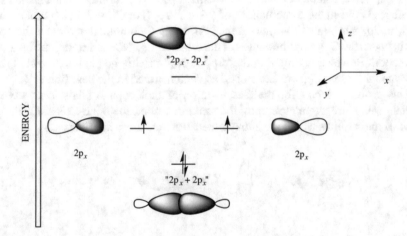

Since 2p orbitals can have three different spatial orientations along three coordinate axes, it is necessary to consider the other two combinations: "$2p_y \pm 2p_y$" and "$2p_z \pm 2p_z$" as well. For convenience let us discuss the "$2p_z \pm 2p_z$" case first.

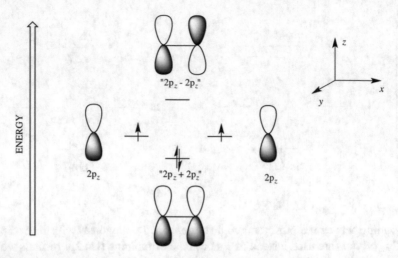

We obtain two molecular orbitals: bonding "$2p_z + 2p_z$" and antibonding "$2p_z-2p_z$". In the same way we can combine "$2p_y \pm 2p_y$" orbitals. What is the main difference between the combinations of ("$2p_z \pm 2p_z$") or ("$2p_y \pm 2p_y$") orbitals and ("$2p_x \pm 2p_x$") orbitals? By rotating the "$2p_x \pm 2p_x$" combination around the bond axis between the C-atoms, the phases of the orbital remain the same, they are invariant to rotation. However, by rotating the "$2p_z \pm 2p_z$" orbital the phases (depicted with gray and white shading) are exchanged as demonstrated in the next figure.

This property serves for the classification of molecular orbitals in two types: orbitals which are symmetric upon the rotation around the bond-axis are called σ-**orbitals** and orbitals which are antisymmetric are called π-**orbitals**.

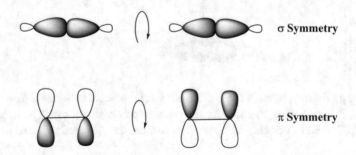

The same reasoning applies to the antibonding combinations. The antibonding orbitals are labeled with symbols σ* and π*.

The extent of stabilization of bonding orbitals and the extent of destabilization of antibonding orbitals both depend on the degree of the **overlap** between atomic orbitals. In σ-orbitals the overlap between $2p_x$ and $2p_x$ atomic orbitals is larger than the overlap between $2p_z$ and $2p_z$ in π-orbitals. For this reason, σ-orbitals have lower energy than π-orbitals and the antibonding σ* orbitals have higher energy than the π* orbitals. The high energy of π* orbitals is the reason why molecules with double bonds between C-atoms are chemically more reactive.

This knowledge can help us in understanding the electron distribution in organic molecules. The lowest in energy electrons are in σ-orbitals, followed by π-orbitals, nonbonding **n**-electrons (the electron lone pairs), π* orbitals and the highest energy σ* orbitals. In the electronic ground states σ, π and n orbitals are occupied by electrons while π* and σ* orbitals are vacant. The nonbonding **n**-orbital in principle does not contribute to the chemical bond, but it is also the cause for the high reactivity of the molecule. All the molecular orbitals are shown in the following figure. However, two of them have additional labels HOMO, and LUMO orbitals. These two orbitals are called **frontier orbitals**, because they are especially important for analyzing possible chemical reactions. HOMO acronym stands for the **h**ighest **o**ccupied **m**olecular **o**rbital and LUMO for the **l**owest **u**noccupied **m**olecular **o**rbital.

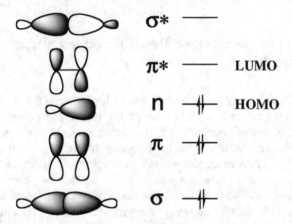

Molecule is in the **ground electronic state** if all valence electrons occupy only bonding or n-orbitals. If the molecule absorbs a quantum of visible or ultraviolet light and if the energy of this quantum corresponds to the energy difference between the bonding and the antibonding orbitals (for instance between π and π*) the electrons can be moved from the bonding to the antibonding orbital and the molecule shall go to the **excited electronic state**. Such excitation is the main characteristic of the reactions induced by light, the photoreactions.

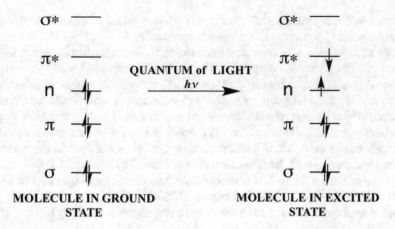

MOLECULE IN GROUND STATE   MOLECULE IN EXCITED STATE

3.3 Distribution of Electron Density, and the Shape of Molecules

Description of the electronic structure of molecules by using this molecular orbital model provides important information about the nature of molecules. One of the most important pieces of information is the electron density distribution. While in σ orbitals the electron density is largest between the atoms i.e., "along the line of the chemical bond" in π orbitals the electron density is concentrated not between the atoms but above and below the plane which contains the "bond line". This is represented in the next figure. In the ethane molecule which possess only σ orbitals, the electron cloud is situated between C-atoms. However, in the ethene molecule which has also π orbital, the electron density is highest above and below the plane in which all the atoms are located. Electron clouds in ethene are not blocked by the atoms and are situated on the open side of the molecule where other particles can attack it. Consequently, ethene molecule is much more chemically reactive then ethane.

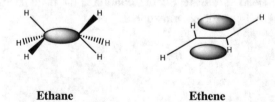

Ethane   Ethene

Although a lot of information about molecular properties can be obtained from the represented qualitative molecular orbital model, for deeper insight into the electronic structure of molecules we need much more sophisticated approach based on the quantitative molecular orbital model. The methods include quantum–mechanical (or to use a better term quantum-chemical) calculations based on the fundamental mathematical and physical concepts. Molecular shape, electron density and charge distribution calculated at such levels provide much more realistic picture of the molecules. In the following figure on the left-hand side, the methane molecule is presented by using simple ball-and-stick model, while on the right hand side the calculated maximum electron density of the same molecule is shown as the green surface.

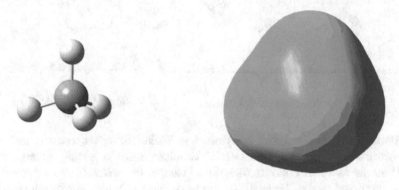

3.4 Bond Lengths, Bond Energies, and Molecular Vibrations

The length of chemical bond depends primarily on the atomic radii of the atoms included in the bond and on the electron interactions. Since in organic chemistry we deal mostly with carbon, hydrogen and a few heteroatoms we will focus predominantly on the length of bonds between these atoms. As it is evident from the table later in this chapter, the single bonds are in principle longer than double bonds, which are in turn longer than triple bonds. Additionally, multiple bonds are stronger than single bonds. However, interatomic distances in chemical bonds are not fixed as implied by the rigid ball-and-sticks models. A much better representation would be a mechanical model in which the balls are interconnected by springs. In such a model the balls are not fixed but are free to vibrate around the equilibrium position. Analogously, the atoms in the molecule vibrate around the equilibrium point at some frequency. The chemical bond length is defined as the interatomic distance when the atoms are in the equilibrium position. Movements of atoms in chemical bonds are nearly hundred times slower than the movements of electrons-waves. To become vibrationally excited, the molecule must absorb electromagnetic radiation of the longer wavelengths than the visible light, thus the radiation in the infrared region.

The simplest molecular vibration is bond stretching. Besides stretching, there are other vibrational modes, for instance vibrations of changing the angle between bonds. Let us analyze the vibrations of water molecule. As it is shown in the next scheme, there are two vibrational modes for bond stretching v_1 and v_2 and one vibrational mode for the angle deformation v_3.

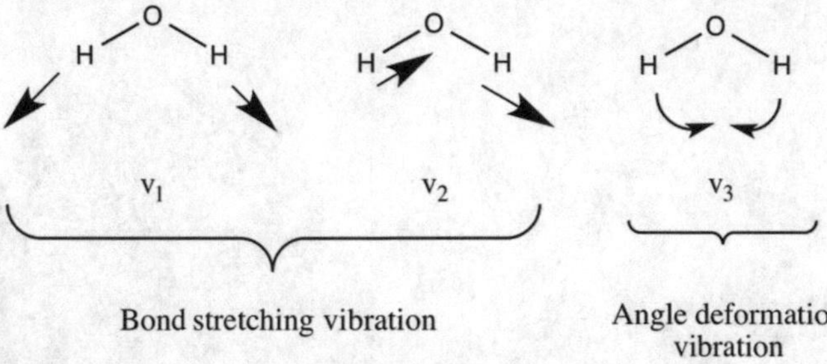

$$v_1 \qquad\qquad v_2 \qquad\qquad\qquad v_3$$

Bond stretching vibration   Angle deformation vibration

Vibrations v_1 and v_2 differ in symmetry. While for v_1 we can say that it is symmetric because both OH bonds stretch simultaneously, v_2 is antisymmetric since the H-atoms move in opposite directions. Hence, the water molecule possesses three vibrational modes. Generally, if the molecule has N atoms it will have 3 N-6 vibrational modes.

Each vibrational mode has its characteristic frequency and the molecule can absorb only the infrared light which has the frequency which correspond to vibrations that are active in this molecule. This fact is used for structural characterization of molecules. Frequencies of the absorption maxima in the infrared spectrum allow us to identify the vibrational modes. Since every functional group possesses characteristic vibrations, the infrared spectrum can serve for identifications of the functional groups and groups of atoms present in the molecule. This method is called **infrared spectroscopy**. In this spectroscopy the vibrational frequencies are represented as wavenumbers which are measured in units called reciprocal centimeters cm^{-1}. The reciprocal centimeter unit is the number of waves within 1 cm. Some characteristic vibrational wavenumbers for most important groups of atoms are presented in the following table.

Group of atoms (functional group)	Wavenumber in cm^{-1}	Mode of vibration
CH_3, CH_2, CH, (alkanes)	2830–2990	C–H stretch
$=CH_2$, CH (alkenes)	3000–3100	C–H stretch
CH_3, CH_2, CH, (alkanes)	1300–1450	HCH bending
C= C, (alkenes)	1600–1660	C=C stretch
C≡C, (alkynes)	2000–2150	C≡C stretch
C= O, (aldehydes, ketones, carboxylic acids, esters)	1700–1780	C=O stretch
C–O, (alcohols, ethers)	1150–1280	C–O stretch
C–Cl, (alkyl-chlorides)	420–700	C–Cl stretch
O–H (alcohols)	3200–3500	O–H stretch

(continued)

(continued)

Group of atoms (functional group)	Wavenumber in cm^{-1}	Mode of vibration
N–H (amines, amides)	3300–3600	N–H stretch

In the figure below is shown the infrared spectrum of acetone, CH_3COCH_3.

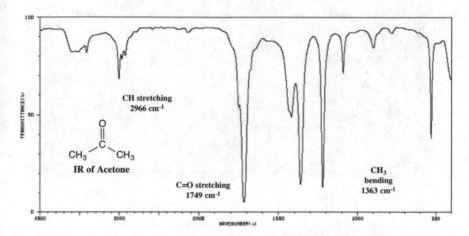

Signals at 2966 cm^{-1} are assigned to the CH stretching of the methyl groups, the sharp maximum at 1749 cm^{-1} is the characteristic stretching of the CO group and peaks at 1363 cm^{-1} correspond to the change of angle between the CH bonds in the methyl groups (bending).

Assignment of all the peaks in spectrum is difficult because the vibrations of acetone molecule are complex: the molecule has 10 atoms and by using the formula 3 N–6 it must have $3 \times 10-6 = 24$ vibrational modes.

Now, we must return to the simplest vibration, the stretching of the bond in a diatomic molecule, for instance, in HCl. Imagine that the H-Cl bond is stretched continuously in one direction. From the mechanical point of view the stretching of the spring increases the potential energy. At some distance the bond will break. Potential energy increases also by compressing the bond. Relationship between the potential energy and bond length is represented in the next figure. Such curves are known as **Morse curves**. As it is shown in the following figure, at the interatomic distances larger than a certain value, the potential energy becomes constant because the chemical bond is broken. The distance at which the potential energy has minimum value is associated with the length of the chemical bond. In reality, the molecule is never at this minimum because the molecule constantly vibrates and the energy of this vibration must be larger than this minimum. The difference between the minimum energy and the lowest vibrational energy level is called **zero-point energy** and it depends on the frequency of vibration by the formula $E = \frac{1}{2}h\nu$ where h is **Planck constant** and ν is the vibrational frequency. Generally, if the molecule is cooled down to the temperature of absolute zero than all its vibrations will have energy $\frac{1}{2}(h\nu)$.

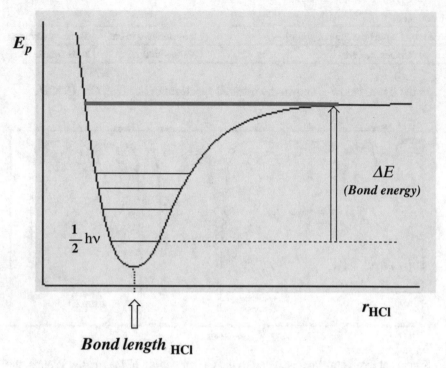

Bond length $_{HCl}$

In analogy with the energy of electrons, the energy of molecular vibration cannot have any value, but is distributed amongst the accessible quantum energy levels. Such energy levels have defined values and the distance between them diminishes as the potential energy increases. If the energy reaches ΔE the potential energy becomes continuous (see the previous figure) and the chemical bond is broken. Hence, the energy ΔE required to break the chemical bond is called the **dissociation energy**. There is correlation between the dissociation energy and the **bond energy** i.e., the energy necessary to break the chemical bond. Strength and energy of the chemical bond depends on the energy difference between the zero-point energy, and the energy of continuum. If chemical bond is strong, the potential energy curve is steep and the bond energy (ΔE) is larger, so the position of the energy minimum is at the shorter bond distance. Conversely, weak bonds have wider potential energy curves, small E and the positions of their minima are at large interatomic distances.

Typical values of bond energies and bond distances for the chemical bonds in organic molecules are listed in the following table. Please note, that the values in the table are averages, based on the measurements for the large number of molecules. Accurate parameters for individual molecules can be obtained from the high-level quantum chemical calculations or by the use of sophisticated instruments.

Covalent bond	Bond length in pm	Bond energy in kJ/mol
C–H	106	413

(continued)

(continued)

Covalent bond	Bond length in pm	Bond energy in kJ/mol
C–C	154	347
C=C	134	614
C≡C	121	811
C–O	113	351
C=O	116	711
C–Cl	163	326
C–Br	191	276
O–H	94	464
N–H	98	389
S–H	132	339
Cl–H	127	431

Important conclusion from this discussion is the rule that the bond energy and bond length are in inverse relationship: as the bond gets shorter it becomes stronger and *vice versa*, as the bond is weaker its energy is smaller. This is important information, because from the data about bond lengths it is possible to predict which bond is weaker and consequently which is more prone to breaking in chemical reaction.

In principle, by using values of bond lengths such as those represented in the table, it is possible to predict molecular geometry and from this the possible reactivity of molecules. Additionally, the sum of all bond energies in the molecule represents approximate value of the **enthalpy of formation** of the corresponding compound. The enthalpy of formation is the energy which is required for the deconstruction of the molecule in atoms, or, conversely, the loss of energy if these atoms form the molecule. For example, in the molecule of ethanol CH_3CH_2OH we count five C–H bonds, one C–C bond, one O–H bond and one C–O bond so the approximate enthalpy of formation, ΔH_f, is therefore: $5 \times 413 + 347 + 464 + 351 = 3227$ kJ/mol. The enthalpy of formation is the measure of the **stability** of molecule.

3.5 Determination of the Molecular Structure by Nuclear Magnetic Resonance Spectroscopy

Besides infrared spectroscopy the most convenient method for deducing the molecular structure is the **nuclear magnetic resonance spectroscopy** (**NMR**). The basic principle of this method is the detection of changes in the orientation of the atomic nuclear spin of hydrogen (the spin of proton) and carbon (the spin of the nucleus of the ^{13}C isotope). This technique can also be adopted for other nuclei such as ^{15}N, ^{19}F, etc.

Like the electron, proton has also the intrinsic angular momentum (**spin**) which is something akin to the rotation around its proper axis. Since proton is a charged particle, its "rotation" induces magnetic field. Roughly speaking, the proton resembles a small magnet. Let us imagine the model in which we have two such small magnets with the opposite orientations. If the magnets are distant from each other their energies are equal independently on their mutual orientation. However, inside the external magnetic field the proton whose orientation is favored relative to the external field has lower energy and the proton with the opposite orientation has higher energy. This energy difference is shown on the figure below as ΔE.

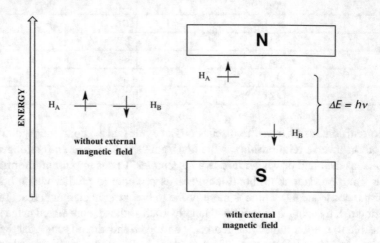

The higher energy proton can change its orientation from disfavored to favored by radiating the quantum of energy which has the frequency v:

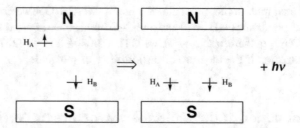

This emitted radiation falls within the radio wave or microwave frequency region. By using the radio wave receiver which can detect this radiation we can detect this quantum transition. However, the transition is so fast that at the moment when the sample is placed into the magnetic field all the protons will reorient themselves so quickly that is becomes very difficult to register signals. To avoid this difficulty the instrument, the NMR spectrometer, has an additional oscillator which irradiates the sample with radio or microwaves by which the spins are constantly reoriented into disfavored orientations. The basic principle of the construction of NMR spectrometer is represented in the next figure.

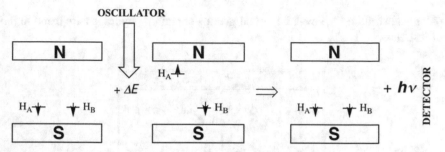

The frequency of radiation depends on the energy difference between nuclei H_A and H_B (middle of the figure above). However, the protons (nuclei of hydrogen atoms) in molecules are surrounded by electron clouds of different densities. Such electron density clouds can reduce the influence of the external magnetic field on the energy required for reorientation of proton spins. We can say that the electrons **shield** the atomic nucleus. The extent of this shielding depends on the functional group to which the hydrogen atom belongs. Thus, the hydrogen in CH_3 group is more shielded than the hydrogen in the CH_2 group. Hydrogen in the $=CH_2$ is shielded even less than the hydrogen in CH_2 group:

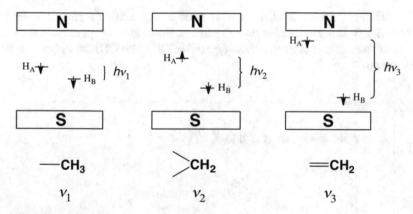

Consequently, in the NMR spectrum of the molecule which contains all three groups (CH_3, CH_2, and $= CH_2$) three signals at v_1, v_2, and v_3, which can be assigned to the three different H-atoms, will appear. This spectroscopy detects hydrogen atoms in different environments in the molecule and is called the **^{1}Hnmr spectroscopy**.

In most NMR spectra the units which are related to the energies of radiation are not frequencies themselves but rather they are differences in frequencies with respect to the corresponding standard substance which has highly shielded protons. Such frequency difference is called the **chemical shift**. In most cases the standard used for measuring chemical shifts is **tetramethylsilane**, $(CH_3)_4Si$ (**TMS**). Frequency differences relative to TMS are small, only millionth parts of the frequencies themselves. Therefore, the chemical shifts are measured in units called **ppm** (part per million).

Chemical shifts for the most important groups containing hydrogen are listed in the table below.

Proton	Chemical shift δ in ppm	Proton	Chemical shift δ in ppm
H–C–R	0,9 - 1,8	$\overset{H}{>}=<$	4,5 - 6,5
H–C–C=C	1,6 - 2,8	H—≡—	2,5
H–C–C̈–	2,1 - 2,5	H–⬡	6,5 - 8,5
H–C–NR	2,2 - 2,9	$H-\overset{O}{\overset{\|}{C}}-$	9 - 10
H–C–Cl	3,1 - 4,2	H–OR	0,5 - 5
H–C–O	3,2 - 3,7	$HO\overset{O}{\overset{\|}{C}}-$	10 - 13

The ^{1}Hnmr spectrum of ethanol constructed from the data for chemical shifts in the table above is represented in the next figure. Since ethanol molecule consists of three different groups containing hydrogen (CH_3, CH_2, and OH) three signals appear in the spectrum.

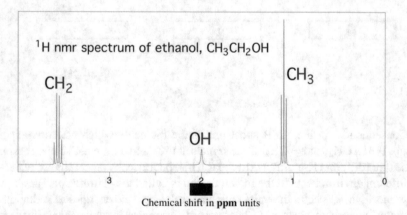

Chemical shift in **ppm** units

The intensities of these signals depend on the number of hydrogen atoms in the corresponding group, so the ratio of intensities of signals is 3: 2: 1 for CH_3, CH_2 and OH. It can be seen from the figure, that the spectrum of ethanol is even more complex. The magnetic moments of hydrogen nuclei are somewhat coupled between each other, because of the influence of the electron cloud. Such coupling between protons is detected from the splitting of spectral lines (four lines for the CH_2 and three lines for the CH_3 group). The splitting of the signal is called the **multiplet**; it depends on the number of hydrogen atoms in the neighboring group. If neighboring group has n hydrogens then the signal is split into $n + 1$ lines. These multiplets are called according to the number of lines: two lines form a **doublet**, three lines **triplet** and four lines **quartet**. In the above spectrum of ethanol, the CH_3 group signal is split into triplet because the neighboring CH_2 group has two hydrogens ($n = 2$), and the CH_2 group is split into quartet because its neighboring CH_3 group has three hydrogens ($n = 3$). Hydrogen in the OH group has special behavior which will be discussed later in this book and its coupling with the neighboring groups is not simple.

The ^{13}C NMR spectroscopy is also important for the determination of the structure of molecules of organic compounds. To obtain the ^{13}C spectrum, the higher outer magnetic field is required, and the instrument must be more sensitive, because the natural abundance of ^{13}C isotope is very low, about 1.109%. The principle of this spectroscopy is the same as for the ^{1}H NMR. As the carbon atoms in molecules are differently crowded with electron clouds, they appear at different chemical shift in the spectrum. Here the scale in ppm (measured relative to TMS) is much more expanded, and goes to 300 ppm. In the following scheme is represented the ^{13}C NMR spectrum of propanal together with the values of chemical shifts for carbon atoms:

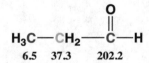

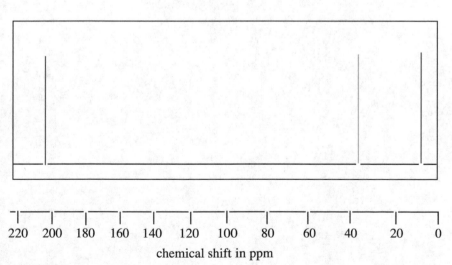

Chapter 4
Alkenes and Alkynes

The simplest alkene is **ethene**, the compound which has previously been called "ethylene". Ethene is produced in large quantities because it is one of the most important substances for production of a wide variety of organic compounds and technological materials. Parent alkyne is **ethyne** which is market known under the trade name "acetylene". This substance is also a very important industrial product. Both ethene and ethyne are flammable; ethyne mixed with oxygen produces a very hot flame so acetylene is used for gas welding. Ethyne can be easily prepared from calcium carbide and water:

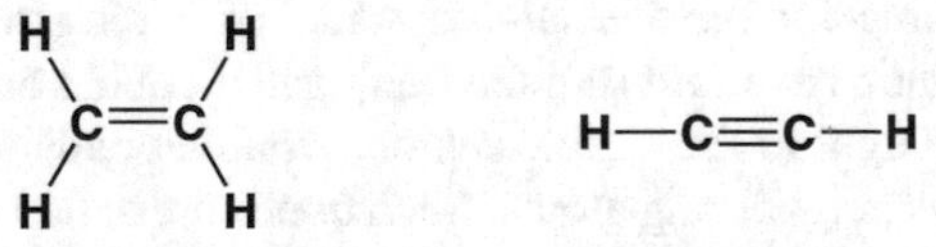

Ethene
(Ethylene)

Ethyne
(Acetylene)

PREPARATION OF ETHYNE

$$CaC_2 + H_2O \longrightarrow C_2H_2 + CaO$$

Calcium carbide   **Ethyne**

4.1 Constitution and Nomenclature

In contrast to the relatively unreactive alkanes, alkenes and alkynes are chemically reactive because they have double and triple carbon–carbon bonds as functional groups. In the IUPAC nomenclature the names of alkenes end with the suffix **–ene** and the position of the double bond is labeled by number of the C-atom on which

© The Author(s), under exclusive license to Springer Nature Switzerland AG 2022
H. Vančik, *Basic Organic Chemistry for the Life Sciences*,
https://doi.org/10.1007/978-3-030-92438-6_4

this double bond begins. In analogy with alkanes with branched chains, the root of the word is the name of the longest chain of C-atoms and the atom numbering is arranged so that C-atom with the double bond has the smallest possible number. The rules for naming alkynes are the same as for alkenes with the exception that the suffix is **–yne**.

$CH_3CH=CHCH_2CH_3$	$CH_3CH=CHCH_2CH_3$
1 2 3 4 5	5 4 3 2 1
Pent-2-ene	**Pent-3-ene**
CORRECT	**INCORRECT**

$HC\equiv CCH_2CH_3$ $CH_3CH_2C\equiv CCH_2CH_3$

But-1-yne   **Hex-3-yne**

simpler:

Butyne

If the alkene molecule has two or more double bonds the names are formed in accordance with their number, for instance **dienes**, **trienes**, **polyenes** etc. If single and double bonds alternate, such alkenes are named **conjugated alkenes**. Such structure is responsible for the special properties of conjugated compounds. For example, alkenes which have four or more conjugated double bonds can be colored because they absorb light in the visible region. Double bonds in the molecule can also be located on the neighboring atoms. Such bonds are called **cumulated** and the corresponding compounds are **cumulenes**. The simplest cumulene is **allene**:

CONJUGATED
DOUBLE BONDS

$H_2C=C=CH_2$ CUMULATED
DOUBLE BONDS
Allene

Cumulenes with ten or more cumulated carbon atoms have been detected in the interstellar space and some theories propose that such molecules played very important role in the formation of organic molecules during chemical evolution.

4.2 Configuration of Alkenes

Since carbon atoms in alkenes are interconnected with double bonds, every C-atom has three neighbors. As we know from previous chapters, the molecule is more stable if groups surrounding certain central atom are positioned as far as possible from each other (VSEPR method). If the central atom is carbon, as in alkenes, its neighbors must lie in the same plane similarly as the C-atoms in graphite. Such planar structure appears not only in alkenes but in all organic molecules that have C-atom with a double bond. In alkynes, carbon atom has only two neighbors and the alkyne molecule is linear. Similar linear structure is characteristic of the structure of cumulene molecules.

The largest electron density in planar molecules like alkenes is not located between the two carbon atoms but rather above and below the molecular plane. Such distribution of electron density is rationalized by the molecular orbital model in which the double bond includes π orbitals as it is shown in the following figure.

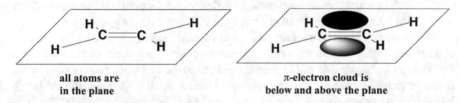

| all atoms are in the plane | π-electron cloud is below and above the plane |

These sterically unhindered π electron clouds are perfect target for the attack of other particles, atoms or molecules. In the previous discussion about the electronic structure of organic molecules it has been mentioned that π orbital is antisymmetric upon the rotation around the bond axis. The consequence of this symmetry property (as well as of electron density distribution) is that torsion of one CR_2 group relative to other CR_2 group causes breaking of the π bond. For example, the structure A* which appears after rotation around the $C = C$ bond in but-2-ene (figure below) has high energy and is very unstable.

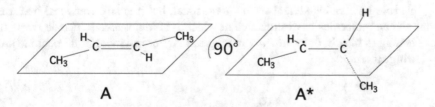

A   A*

Subsequently, A* may experience rotation around the $C = C$ bond and return back to form A or generate a new molecule B:

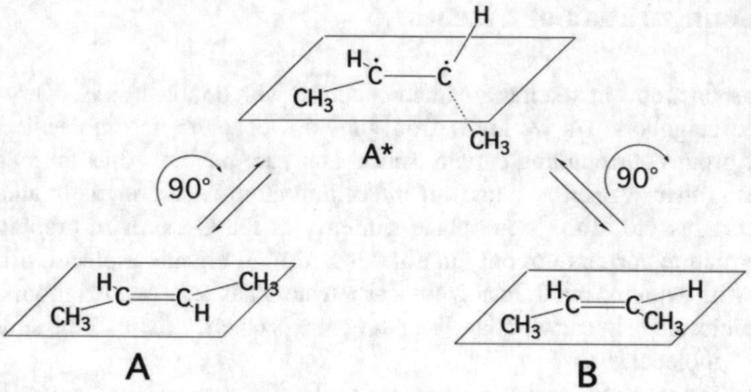

Transformation A $\Rightarrow$ A* $\Rightarrow$ B includes breaking and re-forming of chemical bond (in this case the π bond). Process in which chemical bond is broken or formed is known as **chemical reaction**. If A is transformed into B by a chemical reaction then A and B are isomers. However, these isomers differ only in the spatial arrangement of functional groups; these isomers have different **configurations**. Isomers which differ from each other only in spatial configuration are called **stereoisomers**.

Stereoisomers must be unequivocally named using appropriate chemical nomen-clature. According to the traditional nomenclature such stereoisomers are designated depending on the position of large groups bound to the carbons of the double bond. If the large groups are present on the same side of the double bond, the isomer is labeled by the prefix *cis* and if the groups are situated on opposite sides the prefix is *trans*. We can therefore have *cis*-isomers and *trans*-isomers, respectively.

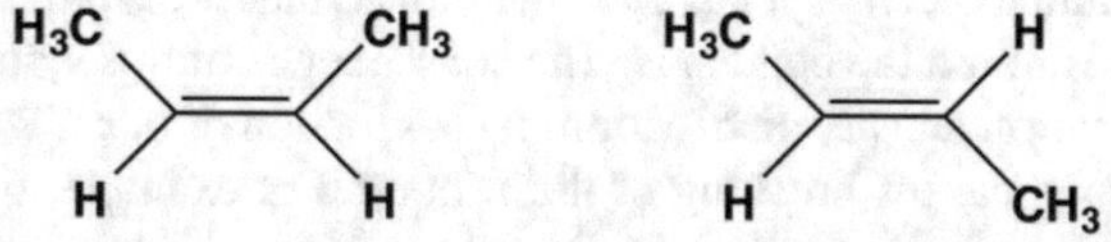

cis-**But-2-ene**   *trans*-**But-2-ene**

This rule for naming stereoisomers is practical, but it is still equivocal because most alkene molecules have structures in which it is impossible to decide whether they belong to the *cis*, or to the *trans* configuration. For instance, let us examine the following structure:

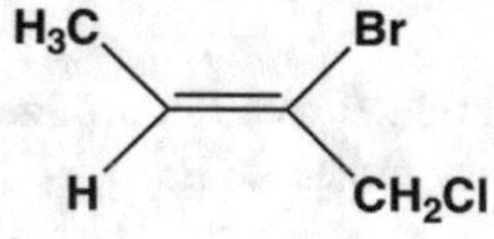

What is the criterion for deciding which groups are larger and which smaller, so as to determine which is *cis* and which is *trans* configuration? To solve this problem

Robert Sidney Cahn, Sir Christopher Ingold and **Vladimir Prelog** have developed the new system of nomenclature for stereoisomers.

Let us apply the **Cahn-Ingold-Prelog** (CUP) system to the molecule shown in the last example via a sequence of steps. In the first step we divide the C-atoms of the double bond by red broken line as shown in the next figure.

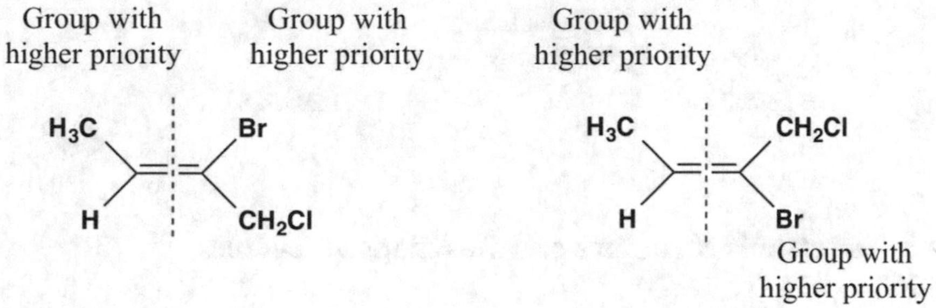

Group with   Group with   Group with
higher priority   higher priority   higher priority

Z-But-2-ene   *E*-But-2-ene

Left and right C-atoms of the double bond are connected to different groups (CH_3 and H on the left C-atom, Br and CH_2Cl on the right C-atom). These groups must be classified by using the **priority rule**. If the groups which have the highest priority are on the same side of the double bond the isomer is labeled with letter *Z* (by the German word *zusammen*), but if the groups with the highest priority are on the opposite sides, the isomer is labeled with letter *E* (by German word *entgegen*).

The priorities depend on the atomic number of the atom bound to the C-atom of the double bond. In our example, the left C-atom is bound to atoms C and H. Since the atomic number of C is greater than the atomic number of H the CH_3 group has priority over H. The right C-atom is connected to atoms Br and C with Br having higher atomic number. Hence, Br has priority over CH_2Cl group. While in the left isomer, the groups with highest priorities (CH_3 and Br) are on the same side and the structure is the Z-isomer, the groups with the highest priority in the right isomers are on the opposite sides and this is the *E*-isomer. The next example is more complicated because both atoms bound to the C-atom of the double bond are the same (carbons in CH_2OH and CH_2Cl).

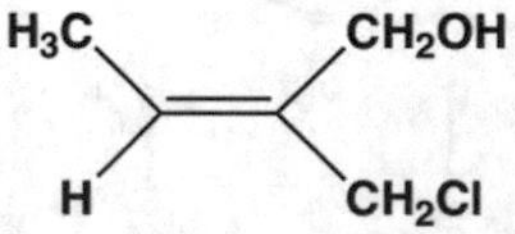

In this case, the priority of $-CH_2OH$ and $-CH_2Cl$ groups must be distinguished by inspecting the atomic numbers of atoms bound to carbons in these two functional groups. Carbon atom in $-CH_2OH$ is bound to H, H and O while carbon in $-CH_2Cl$ is bound to H, H and Cl. Since Cl has larger atomic number than O the $-CH_2Cl$ group has priority over the $-CH_2OH$ group and the configurations are:

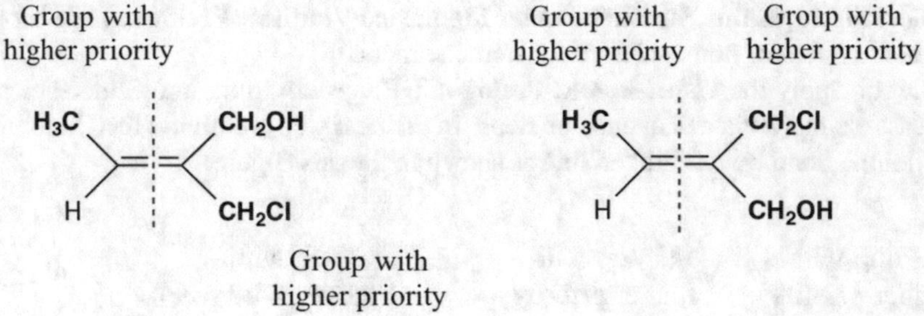

E-isomer   Z-isomer

4.3   Electronic Structure and Reactions of Alkenes

The simplest reaction of alkenes the Z - E isomerization, was already mentioned in the previous section. The transformation in which one isomer (the **reactant**) is rearranged to give another isomer (the **product**) by rotation of 180° requires heating to temperatures above 200 °C in most cases.

However, many *cis–trans* ($Z–E$) isomerizations of natural products are known to occur under mild conditions in living cells. These reactions proceed via different mechanism, utilizing light rather than heat. The well-known example is the reaction which occurs in the eye retina. This reaction represents a fundamental step in the process of visual perception.

The molecule of the derivative **retinal** (see figure below) is an alkene with conjugated double bonds. As we have already mentioned the conjugated molecules have the property to absorb visible light. Upon the absorption of photon, the retinal molecule is transformed from $Z−$ to E-isomer. This change in the molecular stereochemistry triggers the signal by which the information about the absorption of photon is transferred to the nervous system.

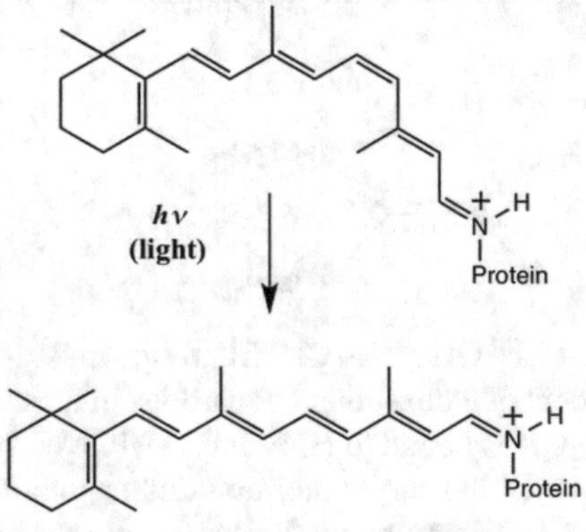

Such photoisomerizations are typical of alkenes and can be explained by shift of one electron from the bonding π to the antibonding π* orbital. This process is shown in the following diagram.

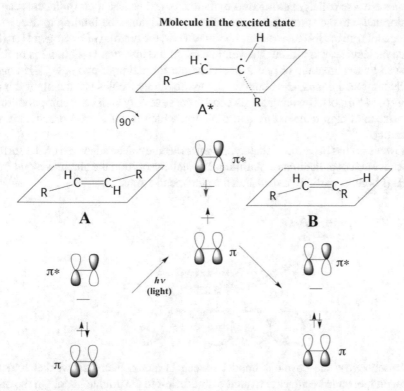

Molecule in which electrons after absorption of photon become shifted to higher orbitals (for instance into π* or σ* orbital) is said to be in an electronic **excited state**. In the example represented in the figure above the excited molecule is described with the structure A*. Since π* orbital has antibonding character as we have already discussed, the double bond in A is broken and in A* has remained only the single σ bond. Because of almost free rotation around the single bond (see chapter about conformations) the molecule can be converted to the new *cis* (or Z) configuration in which all atoms lie in the same plane and the π-bond can be reformed.

Conjugated alkenes have the property that the energy gap between π and π* orbitals diminishes when the number of conjugated double bonds increases. It is the reason why the alkenes with small number of conjugated double bonds absorb ultraviolet radiation while the alkene molecules with large number of double bonds, as it is for instance the molecule of retinal, absorb visible light. Molecules absorbing visible light are colored. Hence, retinal is a good sensor for visible light.

4.4 Addition Reactions of Alkenes, and the Concept of Reaction Mechanism

Alkenes are also called **unsaturated hydrocarbons** because their molecules can be transformed into the **saturated** hydrocarbons, the alkanes, by binding molecules of hydrogen. From the formal structural point of view, the binding of hydrogen H_2 to the alkene molecule occurs through several steps. In the first step H–H bond is broken, followed by the breaking of one π-bond and the formation of two new C–H bonds. Of all mentioned bonds, H–H bond in the hydrogen molecules has the highest bond energy (434 kJ/mol). Breaking of this bond is possible only under extreme conditions, for instance at high temperature and pressure, which is not always practical in the laboratory.

However, it has been found that hydrogen molecule can be adsorbed on the surface of metals such as palladium or platinum. By such adsorption the electron cloud from the H–H bond is partially redistributed to the metal atoms:

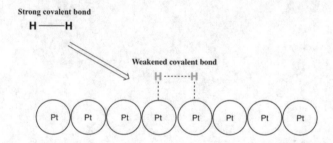

Consequently, the covalent bond between H-atoms becomes weaker and the hydrogen molecule becomes activated for **addition** to the double bond. On the other hand, π-electrons in the alkene molecule have the highest electron density above and below the molecular plane. Thus, the electrons in the alkene molecule are ready for the formation of new C–H bonds with activated hydrogen, as it is shown in the next figure. This type of the reaction is called **addition** and is characteristic for unsaturated hydrocarbons. Addition of hydrogen to the unsaturated compounds is called **hydrogenation**. The role of metal is to activate the hydrogen molecule, therefore, it is the **catalyst**. The reaction can in this case be described as **catalytic hydrogenation**.

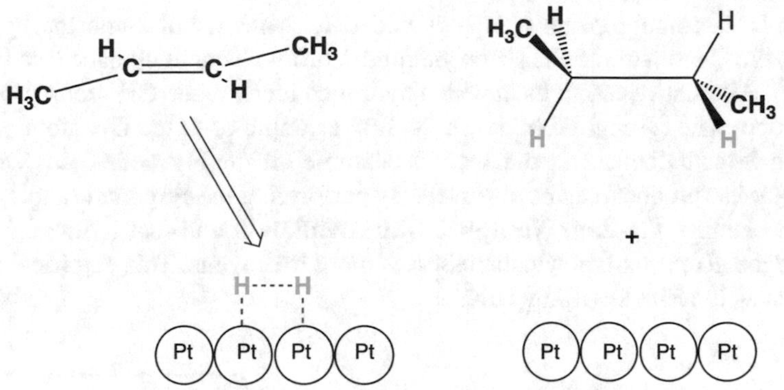

This reaction can be represented by the simple equation:

$$CH_3CH{=}CHCH_3 \xrightarrow{H_2,\ Pt} CH_3CH_2CH_2CH_3$$

The same reaction is also characteristic for alkynes. The only difference is that alkynes have triple bond and the saturation requires two molecules of hydrogen:

$$CH_3C{\equiv}CCH_3 \xrightarrow{2H_2,\ Pt} CH_3CH_2CH_2CH_3$$

Let us imagine that in the first step we add one H_2 molecule to alkyne and the second H_2 molecule is added to the remaining double of alkene, as it is shown in the following scheme. In organic chemistry the addition of hydrogen is called **reduction** while the removing of hydrogen is called **oxidation**. Oxidation also occurs upon addition of oxygen or halogen.

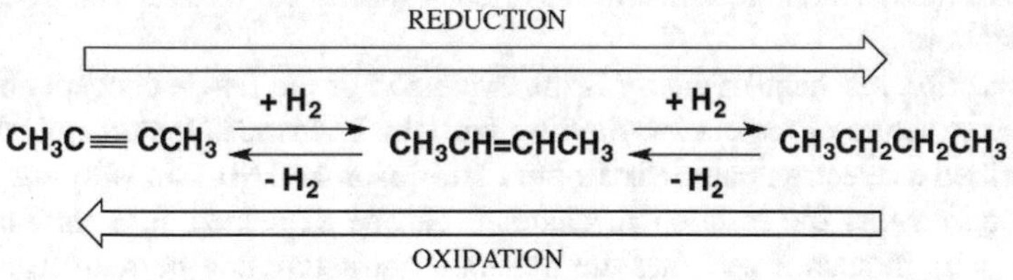

REDUCTION

$$CH_3C{\equiv}CCH_3 \underset{-H_2}{\overset{+H_2}{\rightleftarrows}} CH_3CH{=}CHCH_3 \underset{-H_2}{\overset{+H_2}{\rightleftarrows}} CH_3CH_2CH_2CH_3$$

OXIDATION

4.5 Additions of Hydrogen Halides

Hydrogen halides form alkyl halides by addition to the double bonds of alkenes. For example, hydrogen bromide HBr, can be readily added to propene in the reaction in which hydrogen binds to the carbon atom of the CH_2 group. The resulting product is 2-bromopropane. If HBr is added to 2-methylbut-2-ene the hydrogen binds to the C-atom of the CH group yielding 2-bromo-2-methylbutane (see the next scheme). Alternative reactions in which hydrogen binds to the CH group producing 1-bromopropane (propyl bromide) in the first example or to the C= atom yielding 2-bromo-3-methylbutane in the second example are highly unfavored. Only the traces of these products can be observed. By performing the systematic experiments, Russian chemist **Vladimir Vasiljevič Markovnikov** found that hydrogen always binds to the alkene carbon which possesses more hydrogens. This empirical finding is known as the **Markovnikov rule**.

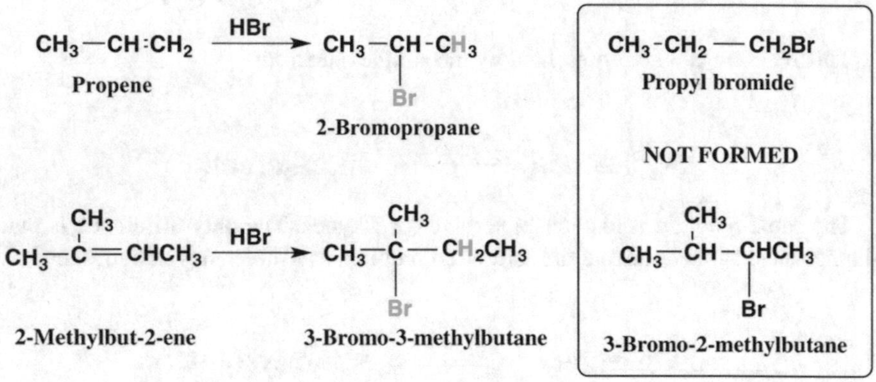

$$CH_3-CH\!:\!CH_2 \xrightarrow{\text{HBr}} CH_3-\underset{\underset{\text{Br}}{|}}{CH}\text{-}CH_3$$

Propene

2-Bromopropane

$$CH_3\text{-}CH_2 \longrightarrow CH_2Br$$

Propyl bromide

NOT FORMED

$$CH_3-\overset{\overset{\text{CH}_3}{|}}{C}=CHCH_3 \xrightarrow{\text{HBr}} CH_3-\overset{\overset{\text{CH}_3}{|}}{\underset{\underset{\text{Br}}{|}}{C}}-CH_2CH_3$$

2-Methylbut-2-ene   **3-Bromo-3-methylbutane**

$$CH_3-\overset{\overset{\text{CH}_3}{|}}{\underset{\underset{\text{Br}}{|}}{CH}}-CHCH_3$$

3-Bromo-2-methylbutane

   Explanation of this rule was not possible in Markovnikov's time because the theory of chemical reactions and electronic structure had not yet been developed. To get deeper insight into the details of the observed behavior, it is necessary to introduce a new chemical concept, the **reaction mechanisms**. The mechanism of chemical reaction describes successive events along the way from the **reactants** (the starting substances) to the **products**. What is the nature of these successive steps? One example of the reaction mechanism was already discussed: the reaction of catalytic hydrogenation.

   The reaction mechanism theory is the extension of the Lewis concept about the electronic structure of molecules. Starting with the Lewis idea that chemical bonds are described as electron pairs, **Sir Robert Robinson** and **Sir Christopher Ingold** have proposed that the reaction mechanism can be explained as a shift of these electron pairs in such a way that the molecular structure can be rearranged from reactants into products.

   Let as apply this concept to the previous examples of additions of hydrogen bromide to 2-methylbut-2-ene. Since the reactant HBr molecule easily dissociates into H^+ and Br^-, the electron deficient H^+ will readily react with the π-electron cloud of the alkene molecule. The H^+ ion is called the **electrophile** because it has affinity for the negatively charged electron clouds. The second ion Br^- is negative and is called the **nucleophile** because it has affinity for the positively charged regions of the molecule of the second reactant. Nucleophiles can be easily recognized because

their molecules always possess the nonbonded electron lone pairs, the **n-electrons**.
Several common nucleophiles and electrophiles are listed below:

$$H_2\ddot{O}: \quad \bar{\,}:\ddot{O}H \quad :\ddot{B}r:\bar{\,} \quad H_2\ddot{S}: \quad \ddot{N}H_3$$

Nucleophiles

$$\overset{+}{H} \quad \overset{+}{Na} \quad H_3\overset{+}{O}$$

Electrophiles

Now we are ready to explain the mechanism of addition of HBr to the alkene
molecule. The electrophile H^+, receipts the electron pair from the double bond and
binds to the carbon atom. In this process the second carbon atom of the double bond
loses one electron and becomes positively charged. Such addition of proton H^+ to
the molecule is called the **protonation**.

$$CH_3-CH=CH_2 \quad \xrightarrow{\overset{+}{H}} \quad CH_3-\overset{+}{CH}-CH_3$$

CARBOCATION

From the upper scheme it is clear that the protonation forms new positively charged
molecule which is called **carbocation** and in which the positive charge is mostly
localized on the carbon atom. Carbocations are relatively unstable molecules, the
reactive intermediates which are persistent only for a short period during chemical
reactions. Since each of the alkene carbons could be protonated two carbocations
can possibly be formed. However, experiments and high-level quantum chemical
calculations have shown that stability of carbocations is different:

$$CH_3-CH:CH_2 \quad \xrightarrow{\overset{+}{H}} \quad CH_3-\overset{+}{CH}-CH_3 \quad \text{or} \quad CH_3-CH_2-\overset{+}{CH_2}$$

stable   unstable
CARBOCATION   CARBOCATION

Since only the stable carbocation can survive long enough to react, the Br^- nucleophile donates its lone pair to the positively charged carbon atom and binds to
it:

$$CH_3-\overset{+}{CH}-CH_3 \quad :\ddot{B}r:\bar{\,} \quad \longrightarrow \quad CH_3-\underset{\underset{Br}{|}}{CH}-CH_3$$

In summary, the reaction mechanism consists of two steps: (i) the protonation of the double bond resulting in the formation of stable carbocation and (ii) the attack of the nucleophile (Br^-) on the positive carbon of the same carbocation.

The stability of carbocation depends on the substituents bound to its positively charged carbon. Amongst hydrocarbons, the most stable carbocations are molecules with the positively charged C-atom surrounded by three nonhydrogen neighbors i.e., the **tertiary carbocations**. Less stable are **secondary carbocations** with **primary carbocations** being the least stable. The simplest examples of carbocations are shown in the following scheme.

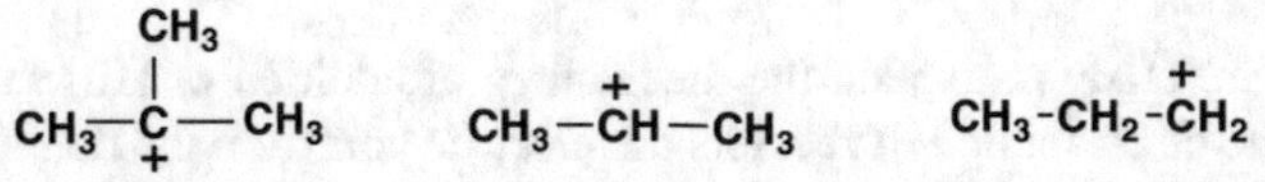

Tertiary CARBOCATION **Secondary CARBOCATION** **Primary CARBOCATION**

MOST STABLE **MOST UNSTABLE**

Important reaction of alkenes is also the addition of water. Since dissociation of water does not produce sufficient concentration of protons required for protonation during the first step in the addition mechanism, sulfuric acid must be added for the efficient reaction. After initial protonation, it follows the nucleophilic attack of the water molecule, with subsequent removal of H^+. Addition of water also occurs in accordance with Markovnikov rule because the preferred reactive intermediate is tertiary carbocation (see the following scheme).

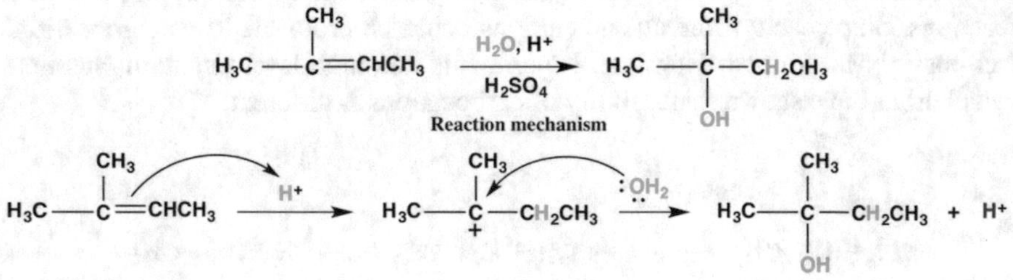

Halogens can also be the reagents for additions to alkenes and alkynes. The products of these additions are dihaloalkanes and tetrahaloalkanes, respectively.

$$CH_3CH = CHCH_3 \xrightarrow{Cl_2} CH_3CH \underset{\underset{Cl}{|}}{} \underset{\underset{Cl}{|}}{} CHCH_3$$

2,3-Dichlorobutane

$$HC \equiv CH \xrightarrow{2\ Br_2} \underset{\underset{Br}{|}}{H \cdot C} \underset{\underset{Br}{|}}{} \overset{\overset{Br}{|}}{C} \cdot H$$

1,1,2,2-Tetrabromomethane

The mechanism of halogen addition is more complex. Since there are no protons in the first step it is necessary that the halogen molecule become **polarized** when being attached to the double bond. Polarization means that the electron density is redistributed between halogen atoms so that one halogen atom becomes partially positive while the other becomes partially negatively charged. In other words, the halogen-halogen covalent bond become polarized. Partially positive halogen atom behaves as electrophile and binds to one of the alkene C-atoms. The carbocation which then appears is a good target for the addition of the remaining negative halogen which acts as a nucleophile. In the scheme below the mechanism of addition of chlorine is shown.

Reaction mechanism

$$CH_3CH = CHCH_3 \xrightarrow{\overset{\delta+}{Cl} \overset{\delta-}{Cl}} \underset{\underset{Cl}{|}}{CH_3CH - CHCH_3} \xrightarrow{:\ddot{Cl}:} \underset{\underset{Cl}{|}\ \underset{Cl}{|}}{CH_3CH - CHCH_3}$$

(δ+ i δ- are labels for partial charges)

From the mechanism of catalytic hydrogenation, we notice that hydrogen adds to the alkene molecule in such a way that both of H-atoms are added to the same side of the molecular plane. Stereochemistry of this reaction is called **syn-addition**. Addition of halogen follows quite different stereochemistry: two halogen atoms approach the alkene molecular plane from the opposite sides. Such stereochemistry is called **anti-addition**. The mechanism is confirmed by the analysis of configurations of products of addition of chlorine to cyclopentene.

anti addition

4.6   Oxidations and Polymerizations of Alkenes

In previous sections we have mentioned that oxidation in organic chemistry includes removal of hydrogen or addition of oxygen. Alkenes can be easily oxidized in reactions with ozone, O_3. Reaction mechanism involves the formation of reactive intermediate the ozonide which subsequently decomposes into two molecules, aldehydes or ketones, depending on the structure of the starting alkene (see the next scheme).

Alkenes can also be oxidized by other reagents such as $KMnO_4$.

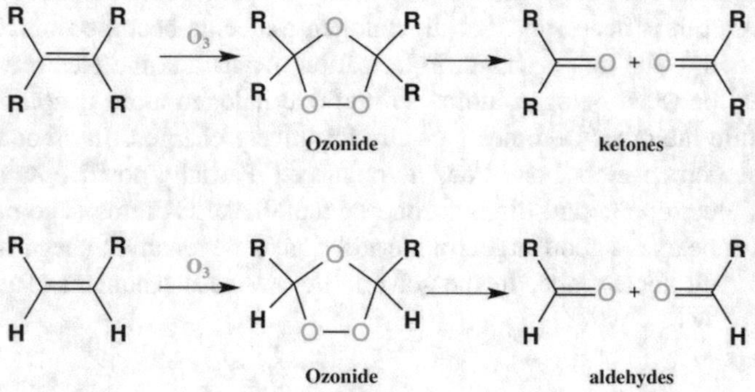

Under high pressure and temperature and in the presence of oxygen or peroxides ethene is transformed into solid elastic substance in which the molecules are connected in long carbon chains. This product is called **polyethylene** and the process is **polymerization**. Polymerizations of alkenes are important processes for the industrial production of different technological materials.

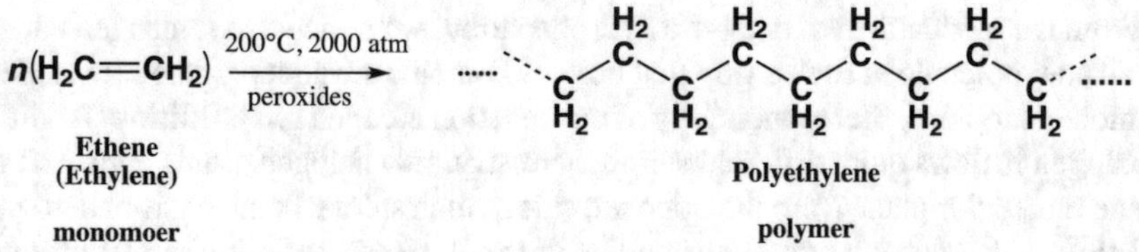

The types and properties of polymeric materials being formed depend on the starting molecule, the **monomer** and on the conditions of polymerization. Almost all industrial polymerizations require the use of appropriate catalyst. Some polymers and starting monomers are listed in the following scheme.

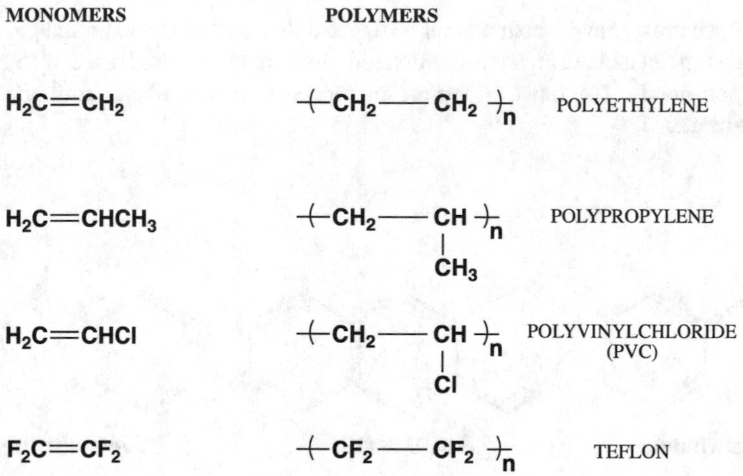

MONOMERS	POLYMERS	
$H_2C{=}CH_2$	$+CH_2{-}CH_2+_n$	POLYETHYLENE
$H_2C{=}CHCH_3$	$+CH_2{-}\overset{\underset{\mid}{CH_3}}{CH}+_n$	POLYPROPYLENE
$H_2C{=}CHCl$	$+CH_2{-}\overset{\underset{\mid}{Cl}}{CH}+_n$	POLYVINYLCHLORIDE (PVC)
$F_2C{=}CF_2$	$+CF_2{-}CF_2+_n$	TEFLON

4.7 Aromatic Hydrocarbons

Although the compounds whose molecules have six-membered ring and conjugated double bonds formally belong to unsaturated hydrocarbons, they exhibit distinct chemical behavior, different from the behavior of alkenes. The simplest compound with such properties is benzene C_6H_6, the substance which has been discovered by **Michael Faraday** two hundred years ago. Derivatives of benzene and other analog compounds were called aromatic compounds because of their characteristic odor. At that time, the determination of the molecular structure of benzene was a big problem for chemists because it has been difficult to deduce its constitution by using the available structural theory. Additions, the most characteristic reactions of alkenes were never observed with benzene and the catalytic hydrogenation of benzene to cyclohexane, C_6H_{12} is possible only under drastic experimental conditions (high pressure and specific catalyst). In the middle of XIX century, **August Kekulé** (the chemist who is together with **Couper** and **Butlerov** known for the discovery of the property of carbon to form chains) proposed that six carbon atoms in the benzene molecule are connected so as to form a six-membered ring:

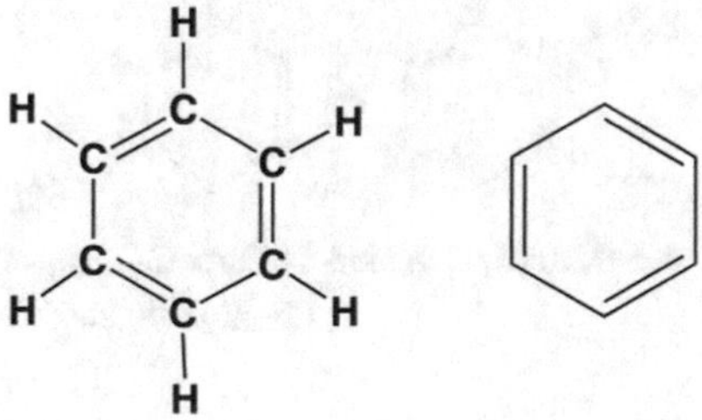

Later, chemists have isolated other more complex aromatic compounds in whose molecules the benzene rings are **condensed** (two rings are condensed if they have a common bond). The most important examples are naphthalene, anthracene and phenanthrene:

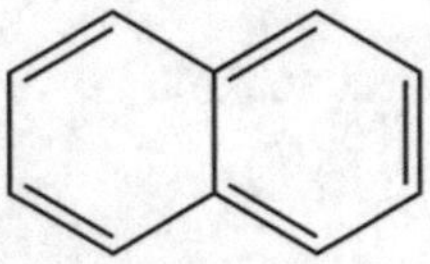

Naphthalene 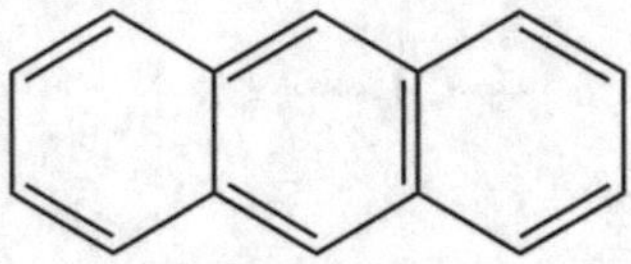   **Anthracene** 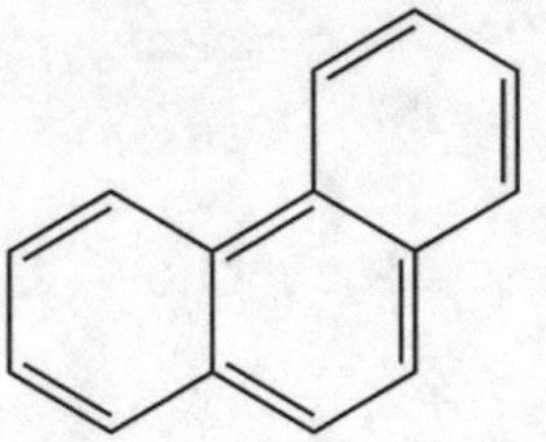   **Phenanthrene**

Aromatic structures can also appear as substituents on the hydrocarbon chains. We then use special prefixes in their nomenclature. For instance, C_6H_5 group is named **phenyl** and $C_6H_5CH_2$ group is called **benzyl** as is shown in the next scheme.

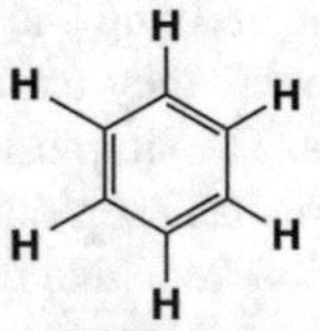

Example:

Phenyl group   **Benzyl group**   **Triphenylmethane**

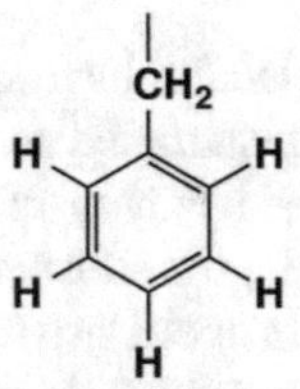

C_6H_5 $C_6H_5CH_2$

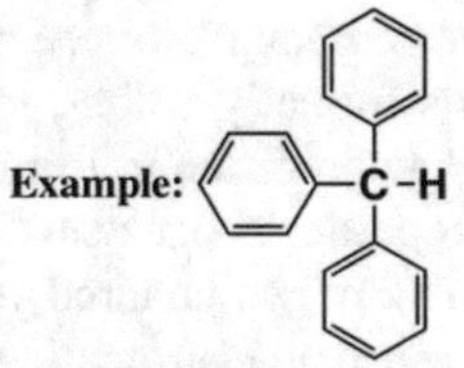

The structure of benzene molecule proposed by Kekulé was not accepted without criticism. The basic concerns were appeared during the investigations of substituted derivates of benzene. For example, if the structure includes the alternating double bonds, as Kekulé has been proposed, the dimethylbenzene derivative should have four isomers:

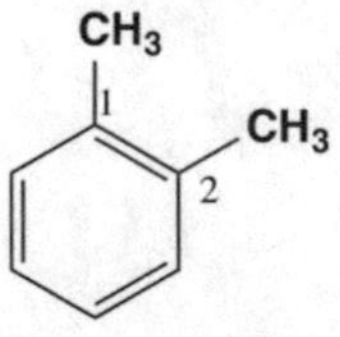

o-Dimethylbenzene   1,6-Dimethylbenzene 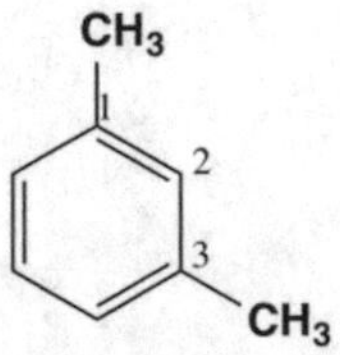   *m*-Dimethylbenzene
1,2-Dimethylbenzene   **1,3-Dimethylbenzene**

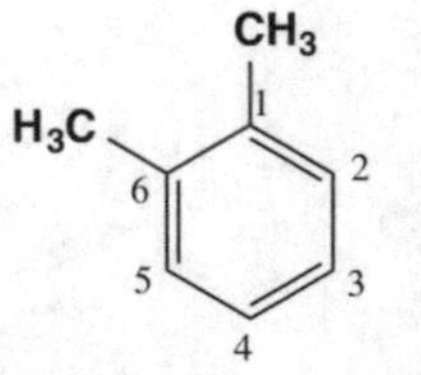

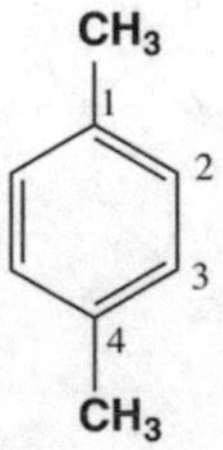

p-Dimethylbenzene
1,4-Dimethylbenzene

The isomers 1,2-dimethylbenzene, 1,3-dimethylbenzene and 1,4-dimethylbenzene also have traditional names: *o*-dimethylbenzene, *m*-dimethylbenzene and *p*-dimethylbenzene (*ortho*-dimethylbenzene, *meta*-dimethylbenzene, and *para*-dimethylbenzene). If the positions of double bonds were **localized**, as proposed Kekulé, 1,2-dimethylbenzene and 1,6-dimethylbenzene would be different isomers because the bond between carbons C_1 and C_2 is double while the bond between C_1 and C_6 would be a single bond. However, all the experiments have demonstrated that only three isomers exist, hence 1,2-dimethylbenzene and 1,6-dimethylbenzene are identical. Consequently, the classical structure theory which implies that the double bonds are localized is unable to describe the structures of benzene molecule and other aromatic compounds.

The resolution of this conflict between the structure theory and experimental results was not possible during the Kekulé's time, because the electronic structure of chemical bond has not yet been discovered.

Only during the first decades of XX century **Robinson** and **Arndt** have proposed the concept of **delocalized electrons**. Six electrons in three formal double bonds are not localized in these bonds but are equally distributed among all six carbon atoms of the benzene ring. This concept was later supported by the discovery that all carbon–carbon bonds in benzene ring are of equally length (139 pm). Benzene C–C bond length is intermediate between the values for single and double bonds (see previous chapter).

Robinson has also proposed two ways to represent the structures with delocalized electrons. First representation is the method of resonance structures which we have already discussed in this book. By using this method, the benzene molecule is represented by two resonance structures. The other representation proposed by Robinson describes the 6 delocalized π electrons by inserting the circle inside the six-membered ring.

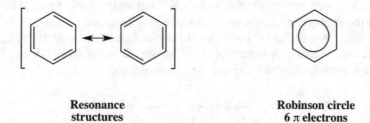

**Resonance
structures**   **Robinson circle
6 π electrons**

Here, it is necessary to point out that the circle corresponds to 6 π-electrons and cannot be used for the descriptions of naphthalene or other aromatics which possess more than six π-electrons. Thus, for the representation of the electronic structures of other aromatic molecules, the method of resonance structures is preferred.

Enumeration of π electrons present in different aromatic molecules reveals certain regularity: 6 π-electrons in benzene, 10 π-electrons in naphthalene and 14 π-electrons in anthracene. The numbers 6, 10 and 14 all result from the formula $4N + 2$, where N is integer. For 6 π-electrons $N = 1$, for 10 π-electrons $N = 2$ and for

14 π-electrons $N = 3$. Compounds whose molecules possess $4N + 2$ delocalized electrons afford greater stability than corresponding conjugated alkenes with planar chains. Such additional stability is called the **aromatic effect**.

Within the molecular orbital theory presented earlier in this book aromatic stabilization is described by π orbitals constructed by combining six $2p_z$ atomic orbitals of carbon atoms which are perpendicular to the molecular plane.

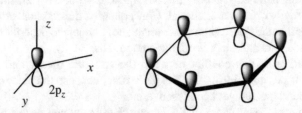

Other aromatic molecules have analogous structures described by the combinations of perpendicular $2p_z$ atomic orbitals. In naphthalene ten $2p_z$ atomic orbitals of carbon are combined into πorbitals.

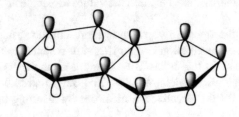

The stabilizing aromatic effect explains why the addition reactions characteristic of alkenes do not occur with aromatic compounds. Addition of chlorine to benzene requires special conditions. The reaction can proceed only by irradiation of the mixture of benzene and chlorine by ultraviolet light. The result of this reaction is a mixture of isomers of hexachlorocyclohexane. Some of these isomers known under the commercial name HCH are used as pesticides.

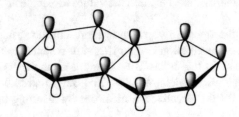

However, their use is restricted because of their negative effect on human health.

Other reactions of aromatic compounds will be discussed later in the section dedicated to electrophilic substitutions.

4.8 Hydrocarbons in Biology

Many unsaturated hydrocarbons are natural products which have important role in the functioning of living organisms and living communities. Relatively high vapor pressure of simple alkenes is responsible for these compounds being used as communication carriers between organisms. Transfer of information must be based on the emission and detection of molecules with particular structure. This effect is known as **chemotaxy** and the molecular carriers of information are called **pheromones**. Some compounds that are pollutants can interfere with pheromones in nature and thus disturb communications between organisms.

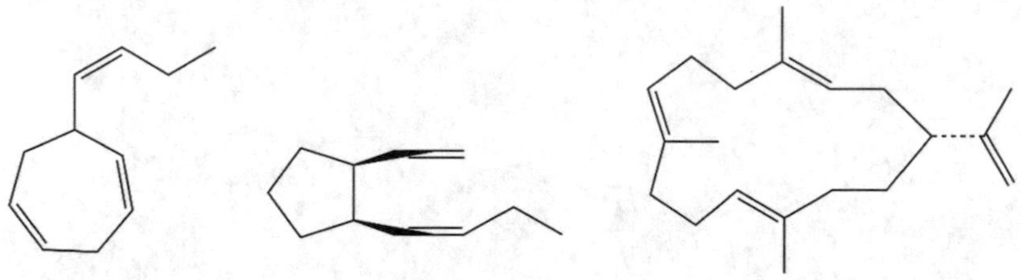

Ectocarpene,   **Multifidene**

Pheromones of brown algae
Phaeophyceae

Cembrane A,
Pheromone of termite secretion

Chapter 5
Substitutions on Saturated Carbon Atom

Alkyl Halides, Alcohols, Thiols, Ethers, Amines

In the previous chapters we have discussed two types of reactions, isomerizations and additions. Another large class of organic chemical reactions is substitution. Substitution in organic molecule can occur either at the saturated or at the unsaturated carbon atom which is called the **reactive center**. The reaction mechanism of substitution includes replacement of the **leaving group** with the **incoming group**.

The incoming groups can be not only nucleophiles or electrophiles which were already discussed in this book, but also other molecular species such as **radicals** which are molecules with unpaired electrons. The general layout of substitution reactions is shown in the scheme below.

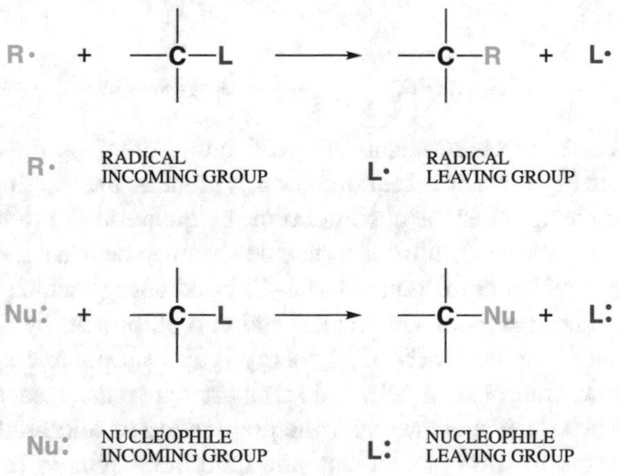

$$R\cdot \;+\; -\overset{|}{\underset{|}{C}}-L \;\longrightarrow\; -\overset{|}{\underset{|}{C}}-R \;+\; L\cdot$$

R·   RADICAL INCOMING GROUP   L·   RADICAL LEAVING GROUP

$$Nu\!:\;+\; -\overset{|}{\underset{|}{C}}-L \;\longrightarrow\; -\overset{|}{\underset{|}{C}}-Nu \;+\; L\!:$$

Nu: NUCLEOPHILE INCOMING GROUP   L: NUCLEOPHILE LEAVING GROUP

© The Author(s), under exclusive license to Springer Nature Switzerland AG 2022
H. Vančik, *Basic Organic Chemistry for the Life Sciences*,
https://doi.org/10.1007/978-3-030-92438-6_5

5.1   Radical Substitutions

5.1.1   Alkyl Halides

The substitution reactions in which the incoming and leaving groups are radicals were perhaps the most important processes in the early phases of chemical evolution. Because radicals are molecules with one unpaired electron, they are exceedingly reactive. In vacuum, radical can have a long lifetime because the probability of encountering another particle and undergoing the reaction is very low. This is why cosmic space and outer layers of planetary atmospheres contain large amounts of radicals. However, even under laboratory conditions some radicals can survive long enough to be detected by spectroscopic methods and be used in practical chemical reactions. Remember that simple inorganic molecules such as NO or NO_2 are also radicals. Some organic radicals are listed in the next scheme.

·CH₃	(CH₃)₃C·	Cl·
Methyl radical	*tert*-Butyl radical	Chlor radical

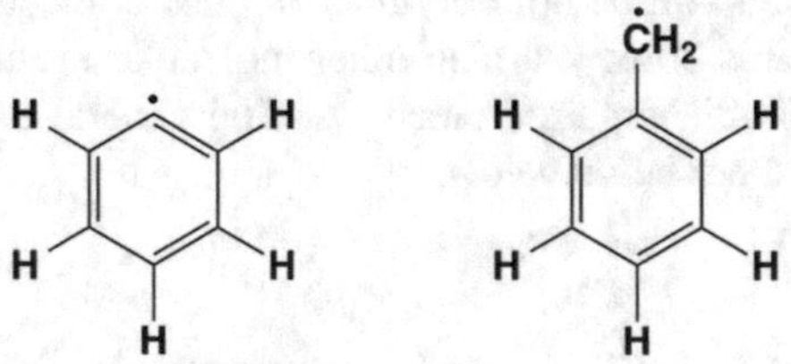

Phenyl radical   Benzyl radical

In the most cases, organic radicals are products of breaking the carbon–carbon single bond. Since high energy is required for this process, the reaction in most cases occurs photochemically after the absorption of the quantum of ultraviolet light. To produce a chlorine radical Cl· chlorine molecule Cl_2 must be irradiated with the light of the wavelength which corresponds to Cl–Cl bond energy which is 243 kJ/mol. Bonds between other halogens are weaker and can be broken by irradiation with light of longer wavelengths. Br-Br bond energy is 193 kJ/mol and I–I bond energy is as low as 151 kJ/mol. This is why halogens are convenient reactants in radical substitution reactions which are used for the preparation of **alkyl halides.**

Irradiation of the mixture of methane and chlorine results in the formation of different chlorides of methane. It has been observed that for this photoreaction to proceed it is not necessary to irradiate the reaction mixture continuously during the reaction. It is sufficient to irradiate it for only a short time and the reaction then will continue in the dark. Since the irradiation is used only for triggering the reaction, this first step is called **initiation**. As it has been explained in the previous section this first step serves only to break the Cl–Cl bond, i.e., to generate two chlorine atoms (radicals). As shown by the equations presented in the next scheme, the chlorine radical removes one H-atom from the methane molecule which results in the formation of HCl and the methyl radical $CH_3\cdot$. Methyl radical reacts in the dark with another chlorine molecule producing methyl chloride and another chlorine radical Cl·. The last two reactions which occur in the dark, they continue spontaneously. This reaction step is called **propagation**. The system of reactions ends with the recombination of the remaining radicals into stable product molecules. This last step is called **termination** and the overall reaction system is called **chain reaction**.

$$Cl_2 \xrightarrow{\ h\nu\ } 2Cl\cdot \qquad\qquad \textbf{INITIATION}$$

- -

$$Cl\cdot \ + \ CH_4 \longrightarrow \cdot CH_3 \ + \ HCl$$

$$\textbf{PROPAGATION}$$

$$\cdot CH_3 \ + \ Cl_2 \longrightarrow CH_3Cl \ + \ Cl\cdot$$

- -

$$2CH_3\cdot \longrightarrow CH_3CH_3 \qquad\qquad \textbf{TERMINATION}$$

In this reaction chlorine radical can also react with methyl chloride abstracting the H-atom:

$$Cl\cdot \ + \ CH_3Cl \longrightarrow \cdot CH_2Cl \ + \ HCl$$

$$\cdot CH_2Cl \ + \ Cl_2 \longrightarrow CH_2Cl_2 \ + \ Cl\cdot$$

Consequently, the final product mixture also contains dichloromethane CH_2Cl_2, trichloromethane $CHCl_3$ and tetrachloromethane CCl_4.

Except for the methyl chloride which is gaseous under standard conditions, the chlorides of methane are liquid substances which are used widely in laboratory, industry and also in medicine. Trichloromethane, $CHCl_3$, is also known under the traditional name of **chloroform** and can be used as anesthetic. Tetrachloromethane or carbon tetrachloride is very good solvent with practical applications in the laboratory, but it must be handled with care because of its carcinogenic properties.

Compounds with different halogens bound to the same carbon atom are known as **Freons** and they are mostly used as gases in the refrigeration systems. Molecules of Freons always contain fluorine and other halogens which are mostly chlorine and bromine. Wide application of Freons is the consequence of their relatively high vapor pressure and low boiling point. In industry, Freons have special nomenclature that is based on the number of carbon, fluorine, and, hydrogen atoms. First number designates the number of C-atoms minus 1, the second number is the number of H-atoms plus 1 and third number gives the number of Cl-atoms. The remaining atoms in the molecule are F-atoms. Some simple Freons and their names are given in the following scheme.

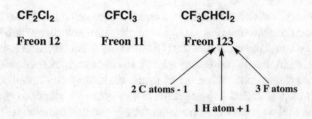

Under the standard atmospheric conditions Freons behave as chemically inert substances, but they undergo photochemical reactions upon irradiation by the ultraviolet light in the higher layers of the stratosphere (at the altitude of 25 km). These photochemical transformations also include reactions with ozone O_3, converting it into the common allotrope of oxygen, O_2. Since the stratospheric ozone layer protects Earth's surface from dangerous ultraviolet radiation, its depletion by reactions with Freons can have negative ecological effects.

The Freon molecule possesses two kinds of chemical bonds, C–F and C–Cl which differ markedly in bond energies. While the C–F bond is strong with the bond energy of 485 kJ/mol the C–Cl bond is much weaker with the bond energy of 331 kJ/mol. Consequently, irradiation of the Freon molecule for instance CF_3Cl, breaks C–Cl bond and yield the radicals:

$$CF_3Cl \xrightarrow{h\nu} Cl\cdot \ + \ \cdot CF_3$$

Very reactive chlorine radical reacts readily with ozone by removing one oxygen atom and forming $ClO\cdot$ radical and the molecule of oxygen:

$$Cl\cdot \ + \ O_3 \longrightarrow \cdot ClO \ + \ O_2$$

Besides ozone and molecular oxygen, the stratosphere also contains oxygen in the atomic form O. Atomic oxygen reacts with $ClO\cdot$ radical regenerating the chlorine radical and one O_2 molecule. The chlorine radical can then react again with another ozone molecule. The chlorine atom regenerated in this process allows a single Freon molecule to destroy millions of molecules of ozone. The overall process is the chain reaction summarized in the scheme:

$$CF_3Cl \xrightarrow{h\nu} Cl\cdot \ + \ \cdot CF_3 \qquad\qquad \textbf{INITIATION}$$

$$Cl\cdot \ + \ O_3 \longrightarrow \cdot ClO \ + \ O_2 \qquad\qquad \textbf{PROPAGATION}$$
$$\cdot ClO \ + \ O \longrightarrow Cl\cdot \ + \ O_2$$

This knowledge has curtailed the production and use of Freons with small molecular masses. However, Freons with higher molecular masses do not undergo such chemical reactions and they are used safely in technological applications, for instance in air conditioners.

The undesirable ecological effects have also restricted the applications of **polychlorinated biphenyls PCB**, the substances which are used for cooling of high voltage electrical transformers. The main disadvantage of the use of PCB is their slow rate of degradation after use. **Dichlorodiphenyltrichloroethane, DDT** is a very efficient insecticide, but its use is restricted because of carcinogenic and neurotoxic effects.

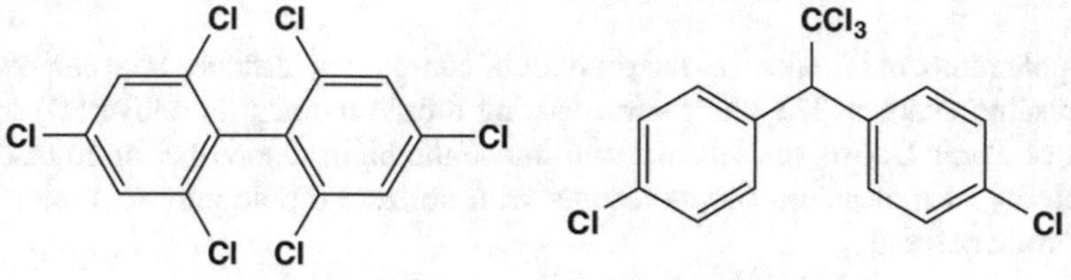

Polychlorinated biphenyl
(PCB)

Dichlorodiphenyltrichloretane
(DDT)

5.1.2 Bond Polarity and the Dipole Moment

As we already know, in the σ covalent bond, the electron density is highest between the two atomic nuclei. However, if the atoms which form the chemical bond are different, the electron density maximum is shifted closer to the atom which has higher **electronegativity**, i.e., the property of an atom to attract electron density. There are different scales of electronegativity, but the most frequently used are values calculated by Linus Pauling. In principle, the electronegativity of an element increases on going from left to right within the period and from bottom to top of the group in the periodic table of elements. For instance, in alkyl halides halogen atom is more electronegative than carbon so that C-atom becomes partially positively charged and the halogen atom partially negatively charged. Partial positive or negative charges are designated in formulas by $\delta+$ and $\delta-$, respectively. The chemical bond in which the charge is unequally distributed is called the **polar bond**. In structural formula the bond polarity is indicated by a vector (bond dipole) as it is shown in the following scheme.

If the molecule has several polar bonds, the average polarity corresponds to the vector sum of bond dipoles that is called the **dipole moment** and labeled by the symbol μ.

**DIPOLE
MOMENT
μ**

Dipole moment is defined as the product of charges and distance between centers of the same charges. The unit for measuring dipole moment is **debye (D)** in the honor of **Peter Debye**, the scientist who made the highest contribution to research on polarity of molecules. The molecules with nonzero dipole moments are called **polar molecules**.

$$\mu = Q \times r$$

(Q charge in Coulombs, r distance between charges in pm)

For the example, if the proton and electron are separated by 100 pm the dipole moment is 4.80 D. This is the limiting value which can be used as a standard for the chemical bond with 100% ionic character [$(1.60 \times 10^{-29}$ C·m)(1D/3.336 $\times$ 10^{-30} C·m) = 4.80 D]. The C–Cl bond-length is 178 pm and the dipole moment is 1.87 D. If we assume that the C–Cl bond is 100% ionic, the expected dipole moment would be $\mu = (178/100)(4.80$ D) = 8.54 D. Since the experimental dipole moment is much smaller, 1.87 D, the character of the C–Cl bond is only 22% ionic (% of ionic character = $(1.87/8.54) \times 100 = 22\%$). In this way the polarity can be related to the ionic character of the chemical bond.

Since the dipole moment is the vector sum of polarities of chemical bonds, it depends on the molecular geometry. Consequently, the molecules such as CO_2 or CCl_4 with symmetrically distributed polar bonds are nonpolar because their total dipole moment equals zero.

As it has been already mentioned, chlorides of methane are very useful solvents. From the knowledge about the polarity of molecules, i.e., their dipole moments we can distinguish between **polar** and **nonpolar** solvents. The polar solvents are for example CH_2Cl_2 and $CHCl_3$, while tetrachloromethane, CCl_4 is a nonpolar solvent. Polarity of solvents is one of the most important properties not only for practical laboratory applications, but also for the theory of reaction mechanisms, reactivity and the selectivity of organic compounds in organic reactions.

5.2 Reactions of Nucleophilic Substitutions and Eliminations

Nucleophiles are electron rich molecules or ions which may contain for example the electron lone pair which is also called the nonbonding electrons. From the building principle of the electronic structure of organic molecules we know that nonbonding n-electrons occupy the highest energy level called the HOMO orbital. Accordingly, these electrons are mostly responsible for the reactivity of organic molecules.

Collision of nucleophile with the alkyl halide molecule will result in a chemical reaction if the nucleophile attaches to the partially positively charged carbon atom. Since the carbon-bromine bond is polarized, the nucleophile will approach the C-atom bound to bromine atom (see following scheme). In the given example the incoming nucleophile is OH^- anion and the leaving group is bromide anion. The product of the reaction of 1-bromopropane with hydroxide anion is propane-1-ol. One functional group (Br) is replaced by another functional group (OH) and the product belongs to the new class of organic compounds – **alcohols**. For simplicity, in the second equation only the nonbonded electrons which are involved in the reaction mechanism are represented.

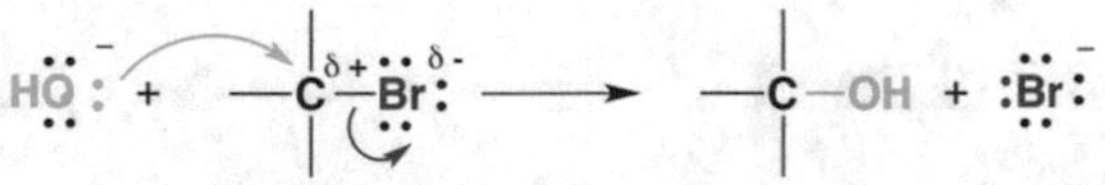

$$HO:^- \; + \quad -\overset{|}{\underset{|}{C}}\!\!-\!\!\overset{\delta+ \;\; \delta-}{Br}: \quad \longrightarrow \quad -\overset{|}{\underset{|}{C}}\!\!-\!OH \; + \; :Br:^-$$

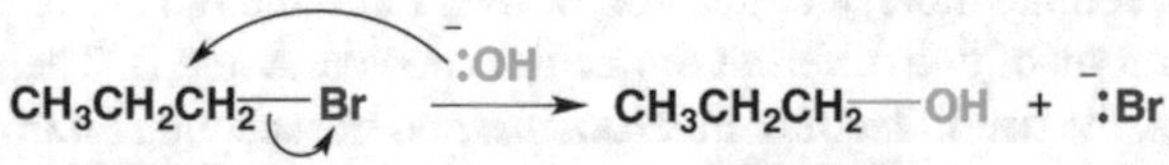

$$CH_3CH_2CH_2\!\!-\!\!Br \quad \xrightarrow{\;\; :OH^- \;\;} \quad CH_3CH_2CH_2\!\!-\!\!OH \; + \; :Br^-$$
1-Bromopropane   1-Propanol

Detailed analysis of this reaction mechanism requires discussion of the electronic structure of both reactants in the framework of the molecular orbital model. We will avoid this approach because it is outside the scope of this book.

The pertinent question is why Br^- as a nucleophile (because it possesses nonbonded electrons) does not react with the product (propan-1-ol) reverting back to 1-bromopropan? Such reversible reaction is not probable because OH^- and Br^- have different **nucleophilicities**. While OH^- is a stronger nucleophile Br^- is a better leaving group. On the basis of experimental experience and high-level quantum–mechanical calculations chemists have quantified nucleophilicity and arranged it into the **nucleophilic order**. In principle, stronger nucleophile replaces poorer nucleophile. Nucleophilicity as well as the nucleophilic order are not universal properties because they strongly depend on the reaction environment, especially on the solvent, but in standard chemical practice the following order applies:

$$^-SH > NH_3 > {}^-OH > I^- > Br^- > Cl^-$$

5.2.1   Reaction Mechanisms of Nucleophilic Substitutions

The systematic measurements of reaction rates of nucleophilic substitutions have shown that depending on the structure of the reactants, the reaction kinetics can follow either the first or the second order rate law. The chemical reaction rate can be expressed by kinetics equations in which the main parameter is the reaction coefficient (rate constant) k:

REACTION RATE = k [A]

      FOR REACTION   **A $\longrightarrow$ products**

REACTION RATE = k [A] [B]

      FOR REACTION   **A + B $\longrightarrow$ products**

While in the first reaction of the scheme above the rate depends only on the concentration of single molecular species A, the rate in the second reaction depends on the concentration of two different molecular species, A and B. Therefore, the first reaction follows the **unimolecular mechanism** and the second reaction is described by the **bimolecular mechanism**.

In most nucleophilic substitutions, the substrate with the leaving group bound to **tertiary** carbon atom follows the **unimolecular** mechanism and the reactant with the leaving group on **primary** carbon proceeds via **bimolecular** reaction mechanism.

Primary C-atom

$$CH_3CH_2CH_2CH_2Cl \xrightarrow{\ \ \overset{\cdot\cdot}{:}OH\ \ } CH_3CH_2CH_2CH_2OH + \ ^{-}:Cl$$

Butyl chloride   Butan-1-ol

BIMOLECULAR KINETICS

Tertiary C-atom

$$CH_3\overset{\overset{\displaystyle CH_3}{|}}{\underset{\underset{\displaystyle CH_3}{|}}{C}}{-}Cl \xrightarrow{\ \ \overset{\cdot\cdot}{:}OH\ \ } CH_3\overset{\overset{\displaystyle CH_3}{|}}{\underset{\underset{\displaystyle CH_3}{|}}{C}}{-}OH + \ ^{-}:Cl$$

2-Chloro-2-methylpropane   2-Methylpropan-2-ol
(*tert*-Butyl chloride)   (*tert*-Butanol)

UNIMOLECULAR KINETICS

Christopher Ingold and **Edward D. Hughes** in the middle of the last century have used such kinetic properties to devise nomenclature for the mechanisms of organic reactions. For the nucleophilic substitutions Ingold and Hughes have proposed labels S_N1, and S_N2, where S and N refer to substitution and nucleophilic, respectively. Numbers **1** or **2** are used to designate unimolecular or bimolecular mechanisms, respectively.

The difference in kinetic behavior between the primary and tertiary substrates is a consequence of different underlying reaction mechanisms. Before discussing details of mechanisms of nucleophilic substitutions, let us mention the basic principles of the theory of rates of chemical reaction established by **Henry Eyring** and **Michael Polanyi**. In a simple chemical reaction, the molecule of the reactant is transformed to the product by passing through high energy structure which is called **transition state**. Hence, the Eyring's concept is known under the name of **transition state theory**. For rearrangement into the transition structure, the reactant molecule must absorb specific amount of energy called the **activation energy**. The lifetime of the transition structure is very short, around 10^{-13} s; this time is comparable to the period of one molecular vibration. In the previous chapters we have already discussed the nature of molecular vibrations and the corresponding spectroscopic techniques, such as IR spectroscopy.

The reaction rate expressed through the rate constant k strongly depends on temperature and on the activation energy. This dependence is expressed in the equation given by **Svante Arrhenius** and **Henricus Jacobus van't Hoff**:

$$k = Ae^{-Ea/KT}$$

where k stands for the reaction rate coefficient, E_a for activation energy, K is **Bolzmann's constant** and T is temperature. If the activation energy is low then the transition structure is more stable, the reaction is fast and consequently the reaction coefficient k is large. This is shown in the next scheme.

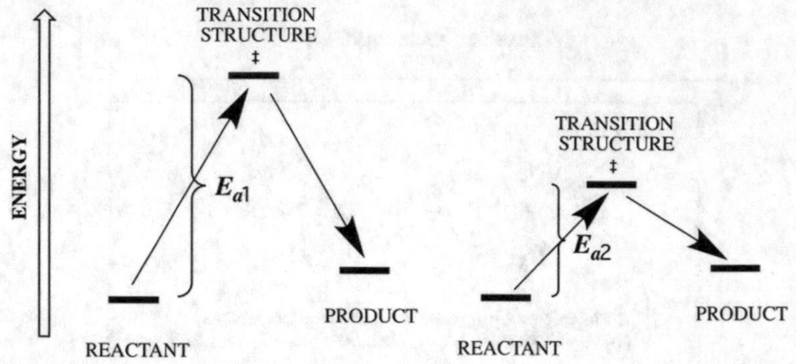

$$E_{a1} > E_{a2} \quad \Rightarrow \quad k_2 > k_1$$

In the two reactions presented, the first one requires higher activation energy than the second one. Consequently, the second reaction is faster. Such scheme describes the reactions that follow the S_N2 mechanism. After the collision of the substrate molecule with the nucleophile, if the molecules have sufficient energy to overcome the activation energy barrier, the transition state is formed. This transition state could be described as a "vibration" in which the incoming nucleophile approaches and the leaving group departs from the reaction center (carbon atom). The breaking of the existing bond and the formation of new bond are concerted processes so S_N2 mechanism is called the concerted mechanism.

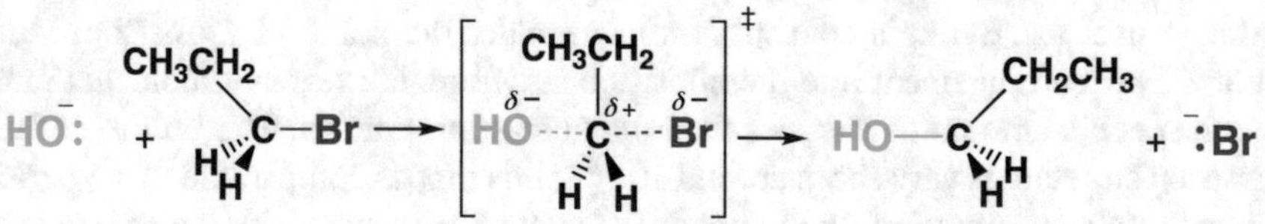

The transition state resembles trigonal bipyramid in which the incoming and leaving groups occupy axial positions and the other three substituents are almost coplanar with the carbon atom. While the central carbon atom holds partial positive charge the leaving and incoming groups carry partial negative charge. The rate of this bimolecular reaction depends on the concentrations of the two molecular species the OH^- ion and 1-brompropane as shown below:

$$\textit{Reaction rate} = k \left[OH^{\bar{:}} \right] \left[\begin{array}{c} CH_3CH_2 \\ \diagdown \\ H^{\backslash\backslash} C - Br \\ H^{\diagup} \;\; H \end{array} \right]$$

Nucleophilic substitutions of tertiary substrates follow more complicated mechanism that consists from two steps. The first step is the dissociation of the bond between carbon atom and the leaving group. The result of this dissociation is the formation of short-lived reactive intermediate called carbocation. In this book we have already discussed carbocations in the section about addition reactions. We have mentioned that the most stable are tertiary carbocations, in which the positive carbon is bound to three neighboring C-atoms. In the substrates such as 1-bromopropane in the last example, only the C–Br bond can be cleaved to form primary carbocation. Since such primary carbocation is highly unstable the reaction proceeds via the concerted S_N2 mechanism in which the primary carbocation does not appear.

Let us discuss the S_N1 mechanism in more detail. Dissociation of the C–Cl bond in 2-chloro-2-methylpropane in the first step yields tertiary carbocation (in this case *tert*-butyl cation) which as a stable species lives long enough to collide with the OH^- nucleophile and form the product. It must be pointed out that the stability of carbocation depends also on the environment. For instance, the cations are more stable in polar than in nonpolar solvents.

Tertiary C-atom

2-Chloro-2-methylpropane   *tert*-Butyl cation   2-Methylpropane-2-ol
(*tert*-Butylchloride)   (*tert*-Butanol)

As it is shown in the scheme above, the reaction consists of two steps: the first step being much slower than the second step. In principle, there are two reactions, one following the other. Such processes are called **consecutive** reactions. The observed rate of such reaction system corresponds to the rate of the slowest step. Such step, in our example the dissociation of the C–Cl bond, is called the **rate-determining step**. Consequently, the observed rate depends only on the concentration of one molecular species, in this case on the concentration of the reactant 2-chloro-2-methylpropane:

$$\textit{Reaction rate} = k \left[\begin{array}{c} CH_3 \\ | \\ H_3C - C - Cl \\ | \\ CH_3 \end{array} \right]$$

The overall S_N1 processes can be described in the framework of transition state theory. Since the first, rate-determining step is slower its activation energy (Ea_1) is higher than the activation energy (Ea_2) for the second, faster step. The complete energy diagram is shown in the next scheme.

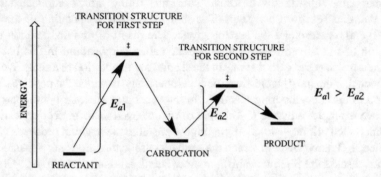

The reaction mechanisms of S_N1 and S_N2 nucleophilic substitutions are convenient for describing reactions where the reactive carbon atom is either primary or tertiary. Reactions of molecules with secondary reactive carbon atom usually follow a mechanism that could be described as the combination of these two mechanisms. The presence of S_N1 or S_N2 mechanism strongly depends on the reaction conditions, mostly on the polarity of the solvent. For instance, polar solvents which stabilize ionic species favor S_N1 mechanism.

5.2.2   Alcohols

In the previous examples we have demonstrated that nucleophilic substitution in which water and hydroxyl group were used as nucleophiles, yields **alcohols** as products. We note that the carbon atom bound to OH group can be primary, secondary or tertiary. Accordingly, alcohols can be classified as primary, secondary or tertiary.

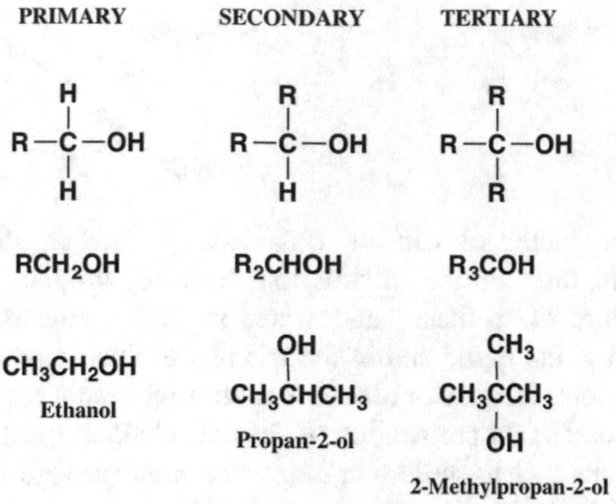

PRIMARY   SECONDARY   TERTIARY

$$
\begin{array}{ccc}
\text{H} & \text{R} & \text{R} \\
| & | & | \\
\text{R}-\text{C}-\text{OH} & \text{R}-\text{C}-\text{OH} & \text{R}-\text{C}-\text{OH} \\
| & | & | \\
\text{H} & \text{H} & \text{R}
\end{array}
$$

RCH_2OH R_2CHOH R_3COH

CH_3CH_2OH

Ethanol

$$
\begin{array}{cc}
\text{OH} & \text{CH}_3 \\
| & | \\
\text{CH}_3\text{CHCH}_3 & \text{CH}_3\text{CCH}_3 \\
 & | \\
 & \text{OH}
\end{array}
$$

Propan-2-ol

2-Methylpropan-2-ol

Alcohols are named starting with the name of the longest carbon chain and using the suffix **–ol**. Numbering of C-atoms must be such that the carbon on which the OH group is attached has the smallest possible number.

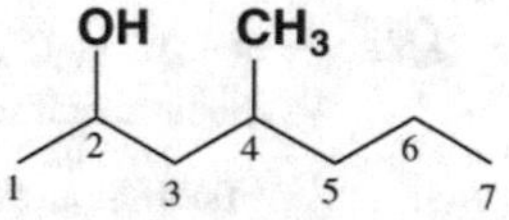

$$
\begin{array}{cc}
\text{OH} & \text{CH}_3
\end{array}
$$

1   2   3   4   5   6   7

4-Methylheptan-2-ol

Alcohols which have two or more OH groups are called **diols**, **triols**, etc. In the following scheme IUPAC names of some alcohols are given together with their trivial names which are still used, especially in trade or manufacturing.

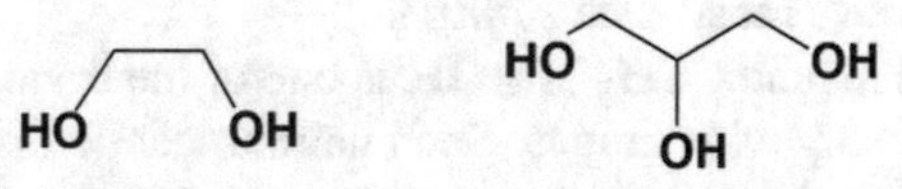

HO   OH   HO   OH

OH

Ethan-1,2-diol   **Propan-1,2,3-triol**
(Ethylene glycol)   **(Glycerol)**

The simplest primary alcohol is **methanol** or **methyl alcohol**, the substance discovered in XVIII century by the Irish chemist **Robert Boyle**. Methanol is one of the products of the dry distillation of wood. It is also the product of different biological processes and appears in some alcoholic drinks, which is illegal because of the poisonous nature of methanol. Although the molecule of methanol is small and has small molecular mass its boiling point, 64.7 °C, is relatively high. Such high boiling point is the consequence of hydrogen bonds between methanol molecules. In this property methanol is similar to water and consequently methanol and water can mix in any amount.

HYDROGEN BOND

HYDROGEN BOND

The 99% pure methanol can be obtained by fraction distillation of the methanol/water mixture. Further distillation does not produce 100% methanol, because the mixture of methanol and water in the volume ratio 99/1 has the same composition in the liquid and in the gas phase. Such mixtures are known as **azeotropes**. Therefore, the pure or **absolute methanol** cannot be obtained by distillation. The best methods for preparation of absolute alcohols must include chemical reactions with metals such as sodium or magnesium. The products of such reactions are compounds that readily react with traces of water.

Reactions of alcohols with metals yield special salts called **alcoholates** or **alkoxides**. For example, methanol reacts with sodium producing **sodium methanolate** (**sodium methoxide**) with the evolution of hydrogen:

$$2CH_3OH + 2\,Na \longrightarrow 2\,CH_3\bar{O}\,Na^+ + H_2$$

Sodium methanolate
or
Sodium methoxide

Sodium methanolate reacts further with water forming sodium hydroxide and methanol. From this solution methanol can be distilled out to get absolute methanol.

$$CH_3O^-Na^+ + H_2O \rightarrow CH_3OH + NaOH$$

For safety reasons, in the preparation of absolute methanol, magnesium is used because it is less reactive metal than sodium.

Ethanol, or **ethyl alcohol**, CH_3CH_2OH is one of the basic products of fermentation of fruits, especially wine grapes. Such natural reaction is known as alcoholic fermentation by which the carbohydrate glucose is decomposed into ethanol and carbon dioxide. The reaction is promoted by special biocatalysts called enzymes:

$$C_6H_{12}O_6 \xrightarrow{\text{Enzymes}} 2\,CH_3CH_2OH + 2CO_2$$

Glucose   Ethanol

Contents of ethanol in wine produced by fermentation cannot exceed 15% because of the large amount of alcohol which deactivates the fermentation enzymes. Distillation of ethanol (boiling point 78.4 °C) from water yields the azeotrope with 96% of alcohol. Absolute ethanol can be prepared by adding alkali metals followed by distillation, similarly to the preparation of methanol. **Propanol** forms two isomers, 1-propanol, $CH_3CH_2CH_2OH$ and propane-2-ol, $CH_3CH(OH)CH_3$. These isomers also have traditional names: *n*-**propanol** and **isopropanol**, respectively. Isopropanol is a very good organic solvent. Both, ethanol and isopropanol are in wide use for disinfection. An excellent liquid for cleaning laboratory glassware is the solution of 5% of potassium hydroxide in isopropanol.

Ethane-1,2-diol or **ethyleneglycol** belongs to alcohols with more than one hydroxyl group. **Ethyleneglycol** is the main component of antifreeze in car engines. Propane-1,2,3-triol is known under the name **glycerol** and is one of the most important alcohols in living organisms. Glycerol is the main structural component of biologically important compounds lipids; this class of compounds also includes fats. Because of its high viscosity and the relaxing effect on human skin, glycerol has wide use in cosmetics. Reaction of glycerol with nitric acid yields glyceryl nitrate, the explosive liquid called **nitroglycerin**. Since pure nitroglycerin is explosive and is so sensitive that cannot be safely transported and handled, its use has been very limited until 1867 when Alfred Nobel discovered that the sensitivity of nitroglycerin can be controlled by the addition of some neutral powders. Such powders which adsorb nitroglycerin are used in explosive called **dynamite**. Taking into account the destructive power of dynamite and its possible military applications, Nobel used profits from his discovery to establish the prizes awarded to people in sciences and arts whose activities promoted world peace and the development of human society.

$$HOCH_2CHCH_2OH + 3HONO_2 \longrightarrow O_2NOCH_2CHCH_2ONO_2 + 3H_2O$$
$$\underset{OH}{|} \qquad\qquad\qquad\qquad \underset{ONO_2}{|}$$

Glycerol   Glycerol trinitrate
Nitroglycerine

Many biologically important compounds are alcohols. Because of their characteristic pleasant odor many alcohols, for the instance **menthol** and **citronellol** are used in the preparation of perfumes. The alcohol with the more complex molecular structure is **cholesterol**, which is the component of cellular membranes and the starting compound for the biosynthesis of various hormones.

Menthol   Citronellol   Cholesterol

Similarity of alcohols and water can be recognized from their behavior in solutions. In XIX century, **Svante Arrhenius** discovered that water molecules in the liquid state dissociate into H^+ and OH^- ions. We know that H^+ ions react with undissociated water molecules and form the oxonium ions, H_3O^+. Analogously, molecules of alcohols can dissociate into H^+ (H_3O^+) and the corresponding alcoholate anion.

$$H_2O \rightleftharpoons H^+ + \ ^-OH$$

$$ROH \rightleftharpoons H^+ + \ ^-OR$$

Example:

$$CH_3\overset{..}{\underset{..}{O}}H \rightleftharpoons CH_3\overset{..}{\underset{..}{O}}{:}^- + H^+$$

Methanol   Methanolate
                      anion

Producing of H^+ ions is associated with the phenomenon of acidity. Here we shall discuss the phenomena of acidity and basicity in more detail.

At the end of XVIII century, **Antoine Lavoisier** argued that acids are compounds that contain oxygen. In recognition of that Lavoisier named the chemical element *oxygen* using the French word *oxygene*, which means acid. However, soon after the work of Lavoisier, Scheele has prepared an acid which does not contain the element oxygen, the hydrochloric acid. After this discovery of hydrochloric acid, the element essential for the property of acidity became hydrogen. At the beginning of XX century, **Johannes Brönsted** and **Thomas M. Lowry** defined acids as substances which are proton donors and bases as proton acceptors.

The equations above showed the behavior of alcohols as acids, i.e., as proton donors. In the presence of strong acids alcohols can also behave as bases, i.e., proton acceptors:

$$R\overset{..}{\underset{..}{O}}H + H^+ \longrightarrow R\overset{+}{\underset{..}{O}}H_2$$

Oxonium ion

PROTONATION REACTION

H^+ ions which originate from the strong acid, bind to the oxygen atom via one of its electron lone pairs. We say that strong acid **protonates** oxygen in the alcohol molecule forming the **oxonium ion**. Accordingly, alcohols can behave either as acids or as bases, depending on the medium.

In general, acidic behavior can be represented by the equation in which the acid is labeled as AH. Anion A^- that remains after removing the proton behaves as a base, because it is in chemical equilibrium and it can again accept the proton. Such base is named the **conjugate base**.

$$AH \quad \overset{K_a}{\rightleftharpoons} \quad \underset{\substack{\text{Conjugate} \\ \text{base}}}{A:^-} \quad + \quad H^+$$

This equilibrium can be characterized by the equilibrium constant K_a. By convention, the equilibrium constant is defined as the quotient of concentrations of products and reactants:

$$K_a = \frac{[A:^-]\ [H^+]}{[AH]}$$

Because the range of equilibrium constants spans many orders of magnitude, it is in practice replaced by the logarithmic parameter pK_a:

$$pK_a = -\log K_a$$

When the acid is more dissociated it is stronger and yields more protons. Hence, if the equilibrium is shifted to right, the constant K_a is larger and the pK_a is smaller. Therefore, strong acids are characterized by small pK_a. Ability of the acid molecule to dissociate depends on the stability of its conjugated base. The alcohol which behaves as a moderate acid is **phenol**, the substance also known as carbolic acid. Because of its acidity, phenol is frequently used as disinfectant in medicine. Dissociation of the phenol molecule forms the conjugate base called phenolate anion:

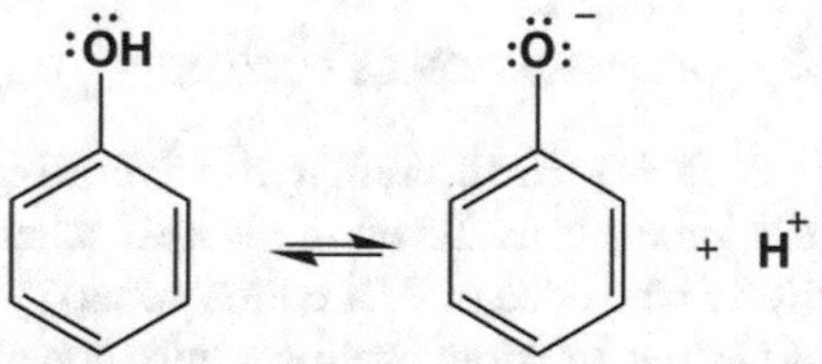

Phenol   Phenolate anion

Stability of the phenolate anion is a consequence of its electron delocalization. Remember that the electron and the molecular energy is reduced by electron delocalization, as we have demonstrated in the case of benzene molecule. The existence of electron delocalization in phenolate ion can be demonstrated using the method of resonance. The resonance structures of the phenolate anion are given below:

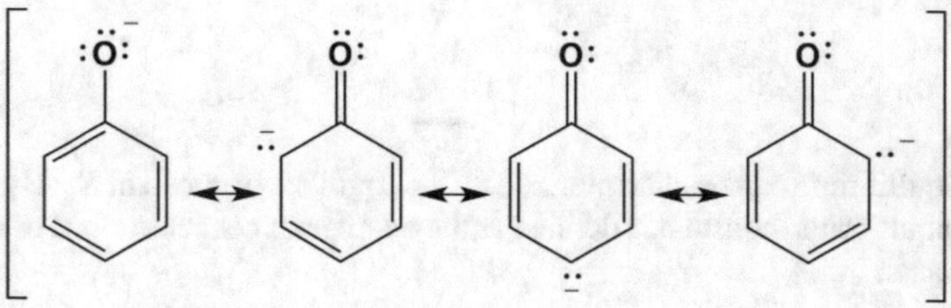

Resonance structures of phenolate anion

As we can see from the formulas above, the negative charge is not localized on the oxygen atom (as is the case in the methanolate anion CH_3O^-) but rather it is delocalized over the entire benzene ring.

The behavior of alcohol as a base is evident in the chemical reaction in which alkenes are formed. In the strong acid medium alcohol is protonated at the oxygen atom, forming the oxonium ion intermediate. The reaction ends with the formation of alkene by removal of the water molecule from this oxonium ion. Such reactions in which some atoms or groups depart from the molecule are called **elimination reactions**. In most cases the preparation of alkenes from alcohols includes reaction with concentrated sulfuric acid, which is a sufficiently strong protonating agent.

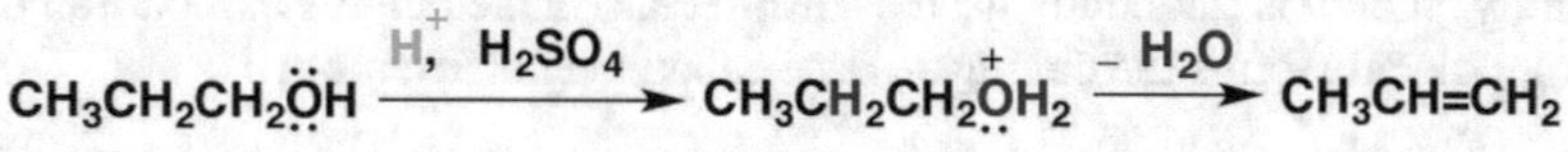

$$CH_3CH_2CH_2\overset{..}{\underset{..}{O}}H \xrightarrow{\;\overset{+}{H},\; H_2SO_4\;} CH_3CH_2CH_2\overset{+}{\underset{..}{O}}H_2 \xrightarrow{\;-\,H_2O\;} CH_3CH=CH_2$$

Propanol   Propene

ELIMINATION REACTION

In the case of tertiary alcohols with different hydrocarbon branches we may expect three different elimination products as shown in the next scheme. However, Russian chemist **Aleksandr Zajtsev** who lived in XIX century used systematic experimental studies to show that out of three possible products only one appears. The other two can be detected only in traces. Zajtsev proposed the rule according to which the preferred product is the one in which the carbons of the double bond have the largest number of alkyl substituents. We say that the reaction obeys **Zajtsev rule**.

$$
\begin{array}{c}
CH_3\ CH_2 \qquad CH_3 \\
\diagdown\qquad\diagup \\
C=C \qquad \text{APPEARS} \\
\diagup\qquad\diagdown \\
CH_3 \qquad CH_3
\end{array}
$$

2,3-Dimethylpentt-2-ene

+

$$
\begin{array}{ccc}
& OH & CH_3 \\
& | & | \\
CH_2\ C & - & CH \\
\diagup & | & \diagdown \\
CH_3 & CH_3 & CH_3
\end{array}
\xrightarrow[\ -\ H_2O]{H,\ H_2SO_4}
$$

2,3-Dimethylpentan-3-ol

$$
\begin{array}{c}
CH_3\ CH \qquad CH_3 \\
\diagdown\diagdown\qquad\diagup \\
C - CH \qquad \text{NOT APPEARS} \\
\diagup\qquad\diagdown \\
CH_3 \qquad CH_3
\end{array}
$$

3,4-Dimethylpent-2-ene

+

$$
\begin{array}{c}
CH_3\ CH_2 \qquad CH_3 \\
\diagdown\qquad\diagup \\
C - CH \qquad \text{NOT APPEARS} \\
\diagdown\diagdown\qquad\diagdown \\
CH_2 \qquad CH_3
\end{array}
$$

2-Ethyl-3-methylbutene

5.2.3 Ethers

Alcoholates (alkoxides) which were already mentioned in the discussion on the reactions of alcohols with metals behave as strong bases and are good nucleophiles. In the middle of XIX century **Alexander William Williamson** succeeded in preparing ethers in the reactions of alkoxides with primary alkyl halides. Later, it has been demonstrated that these reactions follow S_N2 mechanism.

$$
CH_3CH_2\overset{..}{\underset{..}{O}}{:}Na^+ \ +\ CH_3CH_2{-}Br \ \longrightarrow\ CH_3CH_2{-}O{-}CH_2CH_3 \ +\ NaBr
$$

Diethyl ether

Sodium ethanolate
or
Sodium ethoxide

However, the reactions with secondary and tertiary alkyl halides yield alkenes because the elimination reaction is preferred over substitution. In contrast to the previously mentioned elimination of water from alcohol initiated by a strong acid, the **elimination of hydrogen halide** requires a strong base (in this case sodium ethoxide).

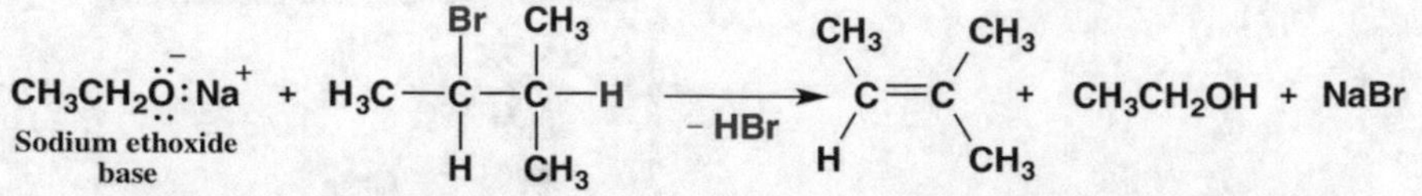

CH3CH2O:Na+
Sodium ethoxide base
Br CH3
H3C—C—C—H
H CH3
- HBr
CH3 CH3
C=C
H CH3
+ CH3CH2OH + NaBr

The base, as proton acceptor, **deprotonates** alkyl halide causing the removal of halide anion in the next step. The reaction proceeds via the transition state in which both C–H and C–Br bonds break **simultaneously**. Such reactions are called **concerted**. The reaction is bimolecular because its rate depends on the concentrations of two molecular species, alkyl-bromide and ethoxy anion. This mechanism is called **E2 elimination mechanism**.

If the leaving group is bound to tertiary carbon atom, the slow reaction step involves dissociation of the C–Br bond and the formation of tertiary carbocation. In that case the reaction rate depends only on the concentration of one molecular species, the alkyl bromide in our example. Accordingly, the reaction is unimolecular and such mechanism is called **E1 elimination**.

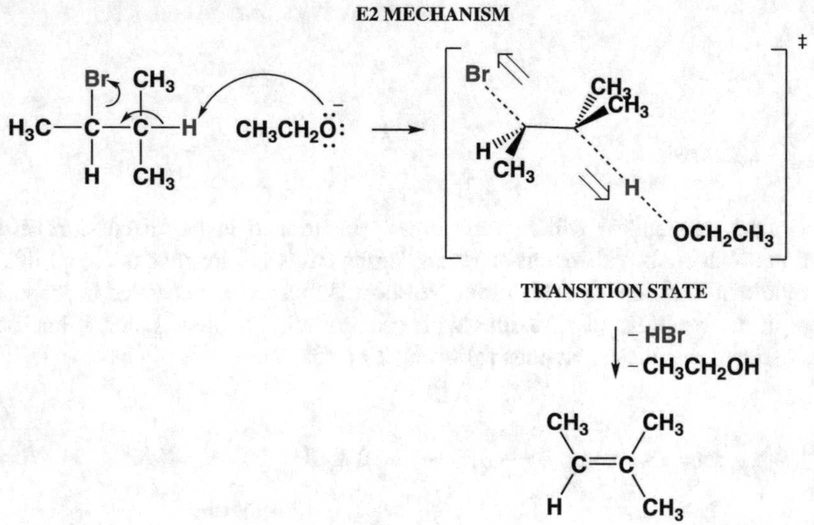

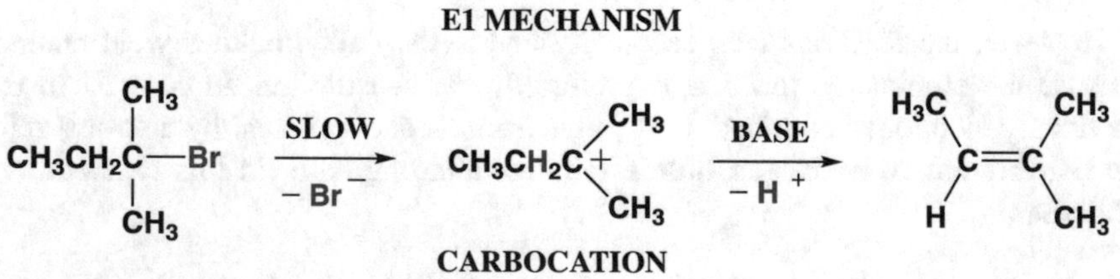

E2 MECHANISM
Br CH3
H3C—C—C—H
H CH3
CH3CH2O:
Br
CH3
CH3
H
CH3
H
OCH2CH3
‡
TRANSITION STATE
- HBr
- CH3CH2OH
CH3 CH3
C=C
H CH3

E1 MECHANISM
CH3
CH3CH2C—Br
CH3
SLOW
- Br−
CH3
CH3CH2C+
CH3
CARBOCATION
BASE
- H+
H3C CH3
C=C
H CH3

Ethers and alcohols can be regarded as structural analogs of water molecule. While in alcohols only one of the H-atoms is replaced with alkyl group, in ethers both hydrogens are replaced with alkyl groups (see the next scheme). In XIX century, Williamson has used such analogy to derive the concept of chemical constitution. This first structural theory was known as the **type theory** in which alcohols and ethers belonged to the water structural type. This was the beginning of the structural formulas which are also used today.

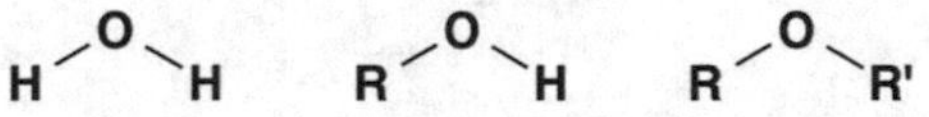

H—O—H   R—O—H   R—O—R'

Water   Alcohol   Ether

STRUCTURAL ANALOGIES

While alcohols have high boiling points because of hydrogen bonds, ethers do not have OH groups and cannot form molecular aggregates. This is why ethers have low boiling points. Dimethyl-ether is gaseous under the standard conditions and diethyl-ether is liquid with the boiling point as low as 34.6 °C. The similarity and analogy between ether molecules and water is also evident in their molecular polarities. Because of their molecular polarity, ethers are good solvents for chemical reactions that occur through polar reaction intermediates. Ethers are partially miscible with water. For instance, diethyl-ether can mix with 8% of water. For this reason, ether as a solvent for reactions must be dried by using standard methods for removing water.

Ether molecules with several ether groups are called oligoethers or polyethers. Some of them have cyclic structures, for instance furan, tetrahydrofuran, pyran or dioxane.

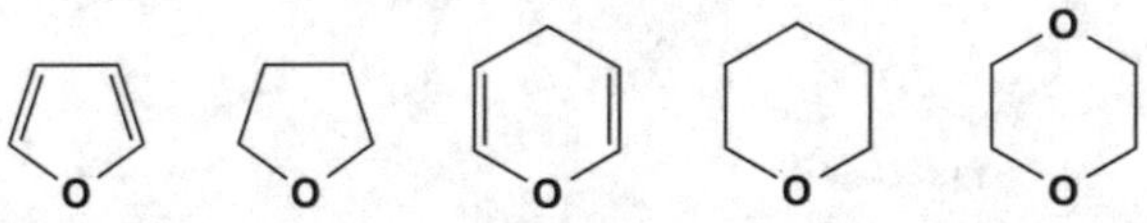

Furan   Tetrhydrofuran   Pyran   Tetrhydropyran   Dioxane

Special group of ethers comprise cyclic ethers with three- and four-membered rings; the epoxides (oxiranes) and oxetanes, respectively. These compounds are very reactive and have important use in industry.

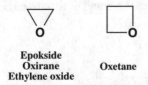

Epokside
Oxirane   Oxetane
Ethylene oxide

Other epoxide compounds are used as starting materials in the production of polymeric materials. The oligomer represented in the following scheme is one of the components of **epoxide resins**.

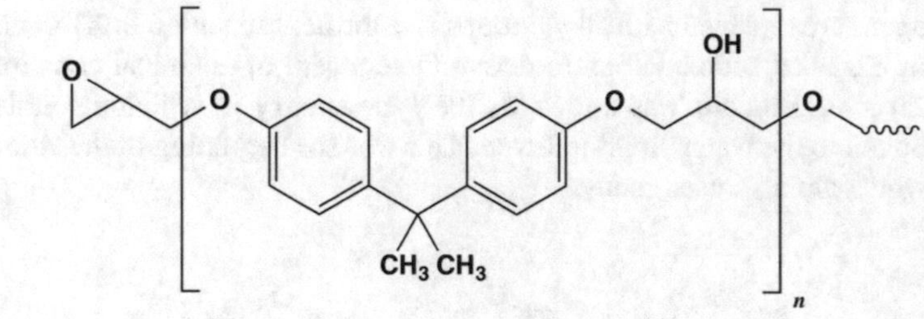

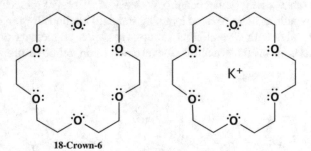

n = 20 - 25 monomeric units

Component of epoxide resin

Cyclic ethers with several oxygen atoms in the ring are of special chemical importance. The simplest, dioxane is routinely used as a solvent in organic synthesis. Ring containing ethers with five or six oxygen atoms afford special chemical properties. Inspection of the structure of the molecules represented in the formulas below shows that oxygen atoms form a cavity which is negatively charged because of the dense electron clouds of electron lone pairs. Such molecules resemble crowns and are known as **crown-ethers**.

18-Crown-6

These ethers are named on the basis of the number of O-atoms and the size of the ring. For instance, the molecule in the scheme above is named **18-chown-6** because the ring has 18 atoms and 6 oxygens. Cavity formed by O-atoms is large enough for metal ions such as K⁺ to enter. Crown ethers with cavities of different sizes can capture metal ions of different ionic radii. Synthesis of crown ethers heralded the rise of the new branch of chemistry, **supramolecular chemistry** which investigates complex molecular aggregates. This will be discussed in more details in last chapters of this book.

5.2.4   Thiols and Sulfides

Sulfur analogs of alcohols and ethers are **thiols** or **mercaptanes** and **sulfides**, respectively. According to the nomenclature rules, thiols are named following the same rules as for alcohols the only difference being that the suffix **–ol** is replaced with **–thiol**. Most of these substances can be recognized by their unpleasant odor. **Ethanethiol** is used as an additive to natural gas since its odor serves as an indicator of gas leak in home installations. Pure natural gas is odorless. Some animals such as skunks release unpleasant odor which originates from thiols shown in the next scheme. However, some thiols have pleasant odor, for instance the components of the aroma of coffee.

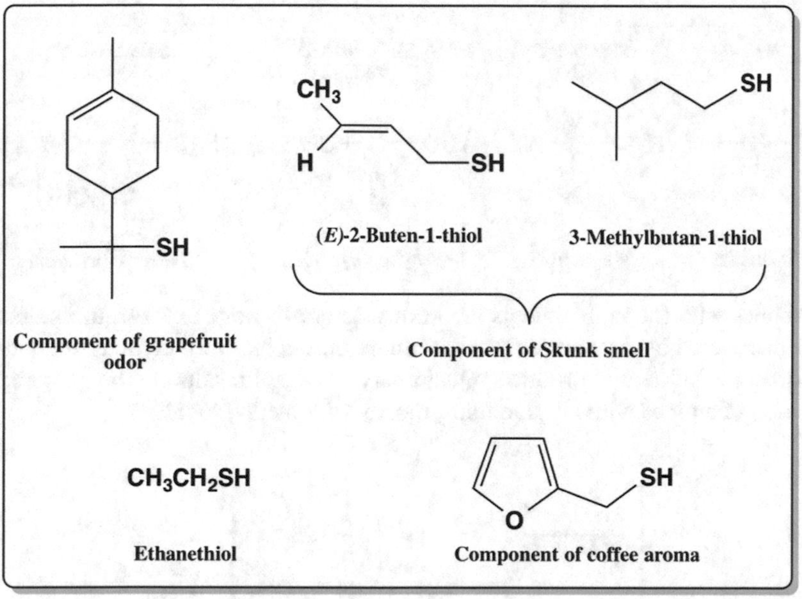

Component of grapefruit odor

(E)-2-Buten-1-thiol

3-Methylbutan-1-thiol

Component of Skunk smell

CH₃CH₂SH

Ethanethiol

Component of coffee aroma

5.2.5 *Amines*

The molecule of ammonia has electron lone pair on the nitrogen atom and is strong nucleophile. In the reaction with alkyl halide, it yields **amines**. The reaction mechanism includes removing hydrogen halide as the leaving group.

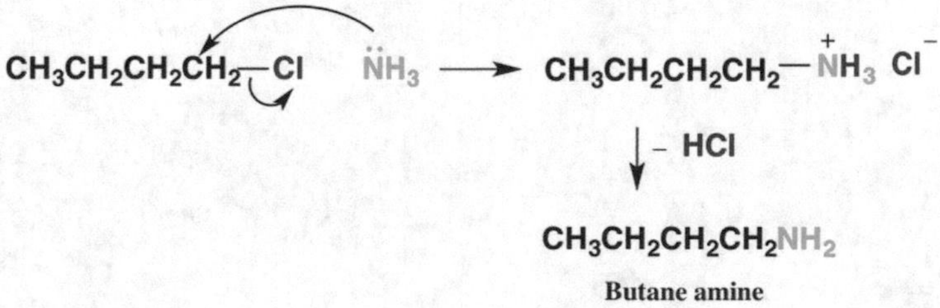

Butane amine

As alcohols and ethers are analogs of water, amines can be considered as analogs of ammonia. Basic chemical properties of amines are consistent with this analogy. Any of the three hydrogen atoms in ammonia can be replaced with alkyl group. Depending on the number of hydrogens substituted, three types of amines can be formed. In the **primary amine** only one H-atom is substituted, in the **secondary amine** two hydrogens are substituted and in **tertiary amine** three hydrogen atoms are replaced with the alkyl group.

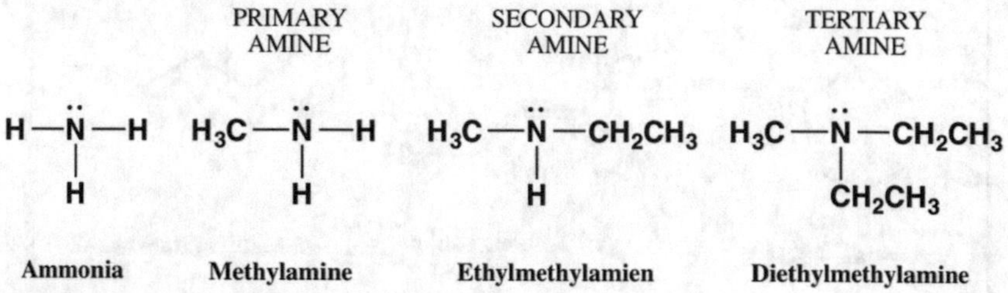

PRIMARY AMINE   SECONDARY AMINE   TERTIARY AMINE

$$H-\overset{..}{\underset{H}{N}}-H \qquad H_3C-\overset{..}{\underset{H}{N}}-H \qquad H_3C-\overset{..}{\underset{H}{N}}-CH_2CH_3 \qquad H_3C-\overset{..}{\underset{CH_2CH_3}{N}}-CH_2CH_3$$

Ammonia   Methylamine   Ethylmethylamien   Diethylmethylamine

Amines with four substituents are known as **quaternary ammonium salts** and were discovered by **Alexander von Hofmann** during his work on determination of the structures of natural products. Quaternary ammonium salts can be prepared by reactions of amines with alkyl iodides, mostly with methyl iodide.

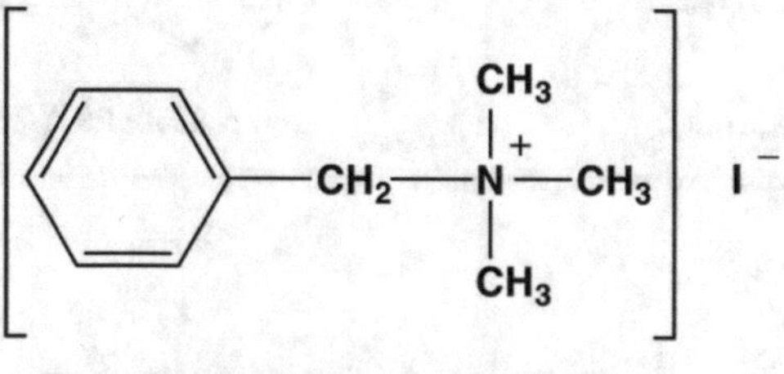

Benzyltrimethylammonium iodide

QUATERNARY AMMONIUM SALT

Since primary amines can be obtained from ammonia by nucleophilic substitution, the secondary amines can be prepared by using primary amine instead of ammonia, while tertiary amines are accessible by starting from secondary amines.

$CH_3CH_2CH_2\overset{..}{N}H_2$ $CH_3CCH_2CH_2{-}Cl$ $\xrightarrow[-\ HCl]{}$ $CH_3CH_2CH_2\overset{..}{N}HCH_2CH_2CH_2CH_3$

Butylpropylamine

$CH_3CH_2\overset{..}{N}HCH_2CH_3$ + CH_3CH_2Cl $\xrightarrow[-\ HCl]{}$

$$
\begin{array}{c}
CH_3CH_2 \\
\diagdown \\
:N{-}CH_2CH_3 \\
\diagup \\
CH_3CH_2
\end{array}
$$

Triethylamine

In water solution ammonia behaves as a base because the molecule can accept proton at its nitrogen lone pair. All amines behave similarly to ammonia: they are **organic bases** because they act as **proton-acceptors**.

$\overset{..}{N}H_3$ + H_2O $\longrightarrow$ NH_4OH

$\overset{..}{N}H_3$ + $\overset{+}{H}$ + $\overset{-}{O}H$ $\longrightarrow$ $\overset{+}{N}H_4\overset{-}{O}H$

$CH_3CH_2\overset{..}{N}H_2$ + H_2O $\longrightarrow$ $CH_3CH_2\overset{+}{N}H_3$ + $\overset{-}{O}H$

Ethyl amine   **Ethylammonium ion**

For example, the protonation of ammonia yields the ammonium ion and the protonation of ethyl amine forms the ethylammonium ion. Behavior of bases can be described by chemical equilibrium and the corresponding constant K_b, similarly as for acids. Analogously, the negative logarithm of K_b (pK_b) is used as a measure of basicity. As pK_b gets smaller, the base is stronger.

$$B: +\ \overset{+}{H} \underset{}{\overset{K_b}{\rightleftharpoons}} \overset{+}{BH} \qquad\qquad K_b = \frac{\left[\overset{+}{BH}\right]}{\left[B:\right]\left[\overset{+}{H}\right]}$$

Conjugate acid

$$pK_b = -\log K_b$$

Acids and bases can also be considered from another point of view. In this view **acids are electron acceptors** while bases are **electron donors**. **G. N. Lewis** has used this **definition** to extend the concept of acidity and basicity to the compounds which don't have hydrogen. The molecule of amine is a stronger base if the electron lone pair on nitrogen is more localized. Localization or delocalization can be estimated

by writing appropriate resonance structures. Electrons are localized if only one resonance formula can be drawn. In principle, if the electron pair is delocalized over a
large part of the molecule, the electron energy is lower, the system is more stable
and does not require the addition of proton.

Let us elaborate this concept in case of basicity of **aniline** and **pyridine**.
Nonbonded electron pair at the nitrogen atom is delocalized over the entire molecule,
as is evident from the resonance structures shown below.

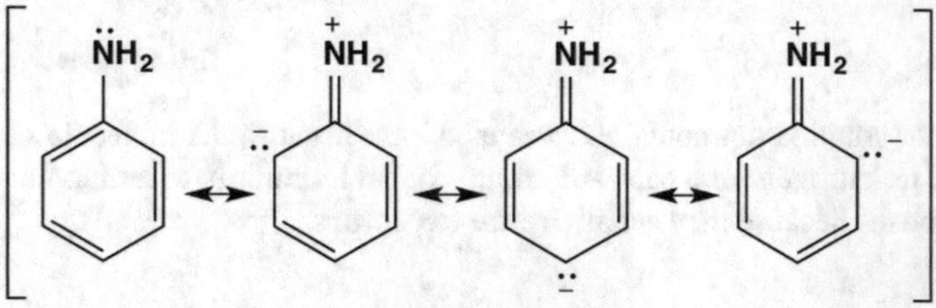

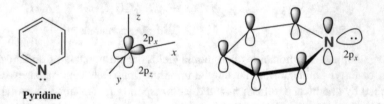

Resonance structures of aniline

In the pyridine molecule the nitrogen electron pair cannot be delocalized because
it occupies the $2p_x$ orbital which is perpendicular to the $2p_z$ orbitals of the aromatic
ring.

Pyridine

Hence, this electron pair remains localized on the nitrogen atom, its electron
energy is higher and its protonation is favored. For this reason, pyridine is stronger
base than aniline.

Chapter 6
Nucleophilic Additions

Compounds with Carbonyl Group

Organic compounds with carbonyl group are not only numerous but also represent excellent substrates for various chemical reactions. This due to the nature of the carbonyl group and its electronic structure. Since carbonyl group has oxygen as heteroatom bound by double bond to the carbon atom, its electrons are distributed through σ, π and n orbitals.

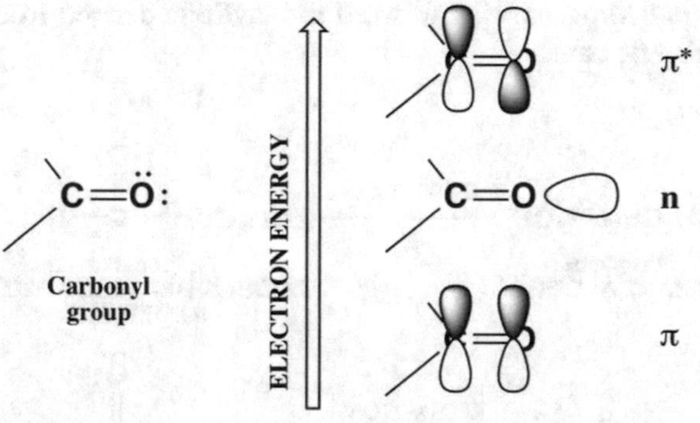

The C=O group is polar because of the high electron affinity of the oxygen atom. Consequently, the carbon atom carries partial positive and the oxygen atom partial negative charge. This polarity is evident in the resonance structures below. The dipole moment of the carbonyl group is relatively high and varies between 1.7 and 3.6 D.

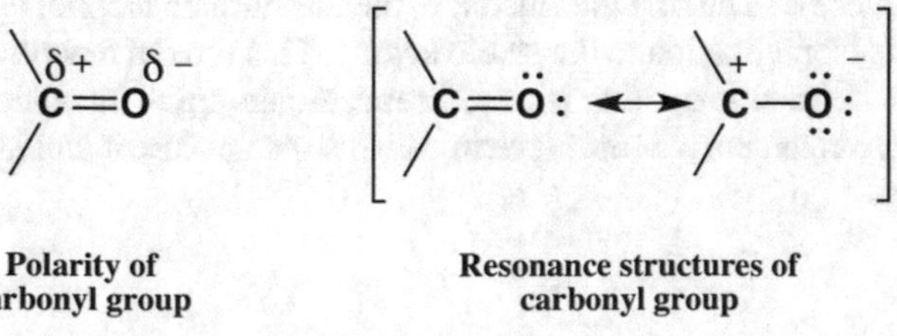

Polarity of carbonyl group

Resonance structures of carbonyl group

© The Author(s), under exclusive license to Springer Nature Switzerland AG 2022
H. Vančik, *Basic Organic Chemistry for the Life Sciences*,
https://doi.org/10.1007/978-3-030-92438-6_6

Nucleophiles can attack the partially positive carbon atom in the carbonyl group (see following scheme). Formation of the new chemical bond between nucleophile and the carbonyl carbon induces the electron shift from the carbon–oxygen double bond to the oxygen atom. The result is the reaction intermediate with the negative charge on oxygen. This mechanism is universal and describes all the reactions called **nucleophilic additions to the carbonyl group**.

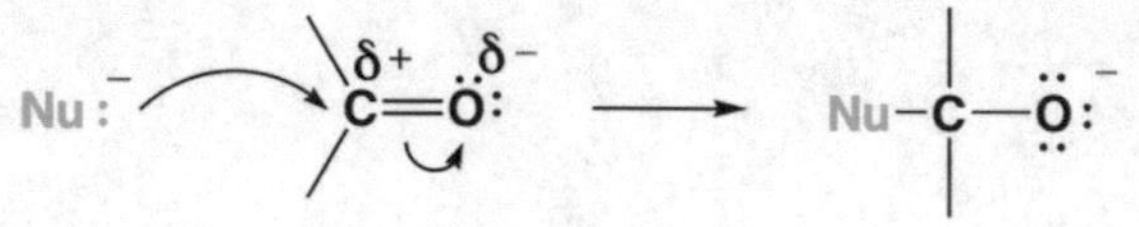

6.1 Aldehydes and Ketones

Aldehydes appear as products of oxidation of primary alcohols while ketones are products of the oxidation of secondary alcohols. Since oxidation is recognized as the reaction in which hydrogen is removed from the molecule, the removal of hydrogen molecule from the primary alcohol yields the compound that has historically been called ***al**cohol **dehyd**rogenatus*. The word **aldehyde** is derived from the first few letters of this historic name.

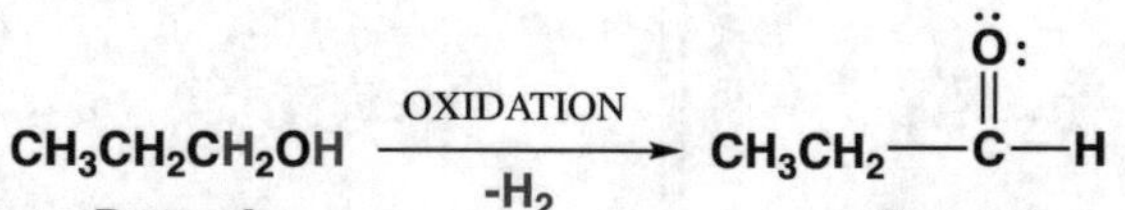

Propanol **PRIMARY ALCOHOL** → **Propanal** **ALCOHOL DEHYDROGENATUS** **ALDEHYDE**

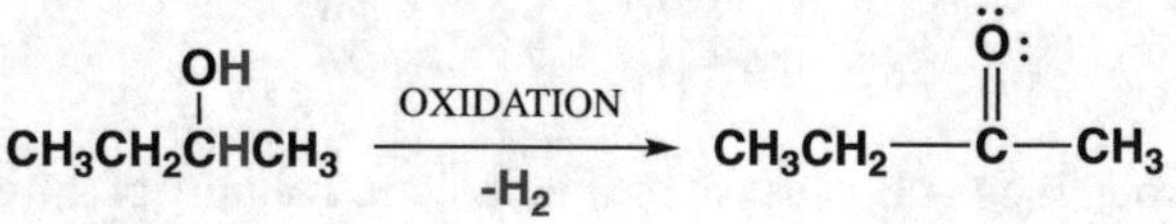

Butan-2-ol **SECONDARY ALCOHOL** → **Butan-2-one** **KETONE**

Besides being prepared by oxidation, aldehydes and ketones can also be prepared by reactions in which the first step includes the addition of water to the triple bond of the alkyne molecule. The first intermediate, the unsaturated alcohol **enol** is unstable and undergoes isomerization to the stable ketone. This type of reaction in which one isomer is transformed to another is called **rearrangement**. The older name for this molecular rearrangement is **tautomerism** and this special case is called the **keto-enol tautomerism**.

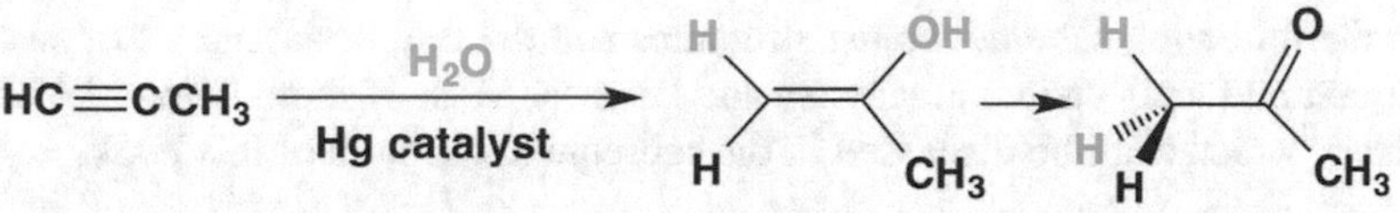

enol   ketone

KETO-ENOL TAUTOMERY

The constitutional difference between aldehydes and ketones lies in the substituents at the carbonyl carbon atom. While in aldehydes one of the substituents is always hydrogen in ketones both substituents are alkyl groups. Therefore, the chemical properties of aldehydes and ketones, especially in nucleophilic addition reactions are similar. The simplest aldehyde is **methanal** or **formaldehyde** and the parent ketone is **propan-2-one** or **acetone**.

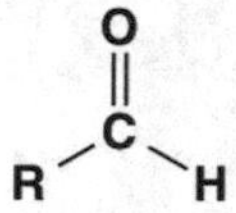

Aldehide 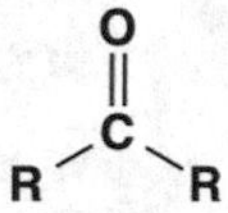   **Ketone** 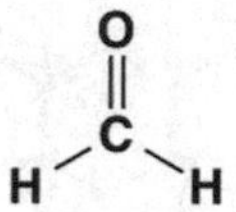   **Methanal**   **Propan-2-one**
                                  (Formaldehyde)   **(2-Oxopropane)**
                                                       (Acetone)

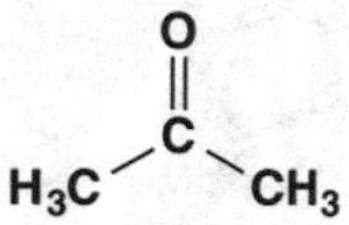

In the systematic nomenclature the names of aldehydes are formed with the suffix **–al**. In complex molecules the aldehyde group is designated by the suffix **–carbaldehyde**. The names of ketones include the suffix **–one** and the numerical label of the carbonyl carbon atom. The numbering of carbons must be such that each one has the smallest possible number. Ketones can also be named by using the prefix **oxo-** and the numerical label of the carbonyl group. The prefix oxo- is used when the compound possesses both aldehyde and ketone functional groups. The longest chain for naming such molecules is defined by the position of the aldehyde, rather than the keto group.

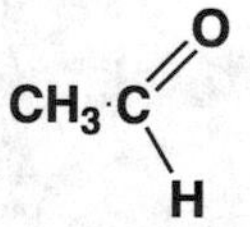

Ethanal   **4,4-Dimethylpentanal** 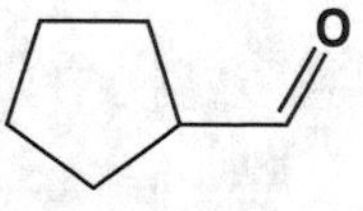   **Cyclopentylcarbaldehide**
(Acetaldehyde)

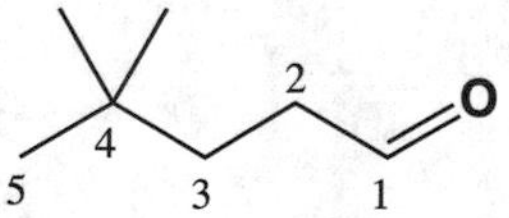

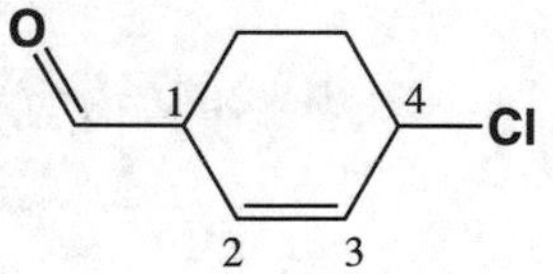

4-Chloro-2-cyclohexencarbaldehyde 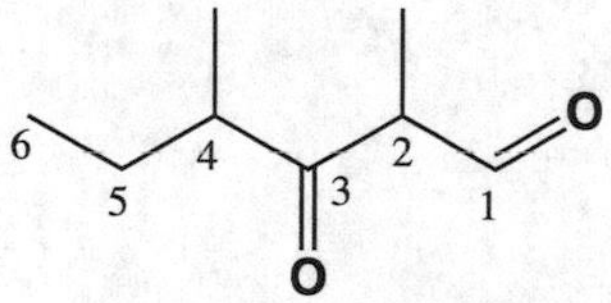   **2,4-Dimethyl-3-oxohexanal**

In the following scheme we list structures and the traditional names of some of important and well-known aldehydes and ketones. Most of them belong to natural products, which shall be discussed in the subsequent chapters of this book.

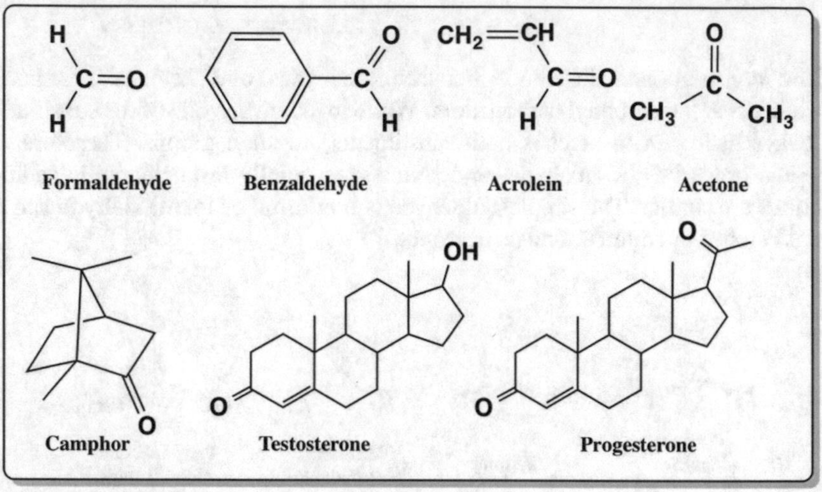

Formaldehyde   Benzaldehyde   Acrolein   Acetone

Camphor   Testosterone   Progesterone

In water solution aldehydes react with water molecules as nucleophiles. The OH⁻ nucleophile attacks the carbonyl carbon forming new C–O bond and shifting the electron pair from the double C= O bond to the oxygen atom. In the reactive intermediate the former carbonyl oxygen becomes negatively charged, as is shown in the next scheme. This negative oxygen is easily protonated and converted to the hydroxyl group. The final product is diol in which both OH groups are bound to the same carbon atom. Such diols are called **geminal diols**. The reaction is reversible and the elimination of water molecule yields the aldehyde reactants.

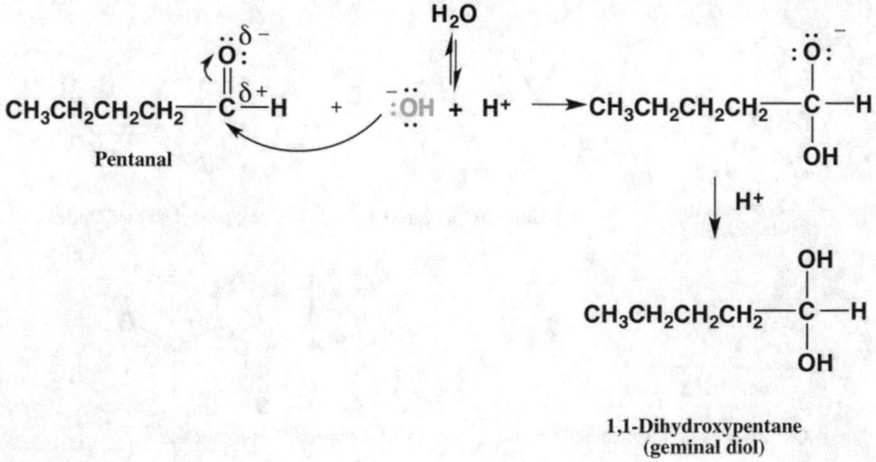

1,1-Dihydroxypentane
(geminal diol)

In general, the reaction can be represented as:

$$
\underset{\text{RCH}}{\overset{\displaystyle \text{O}}{\|}} \quad + \quad H_2O \quad \rightleftharpoons \quad \underset{\text{OH}}{\overset{\displaystyle \text{OH}}{\underset{|}{\text{RCH}}}}
$$

The position of equilibrium depends on the nature of the substituent R. In some aldehydes such as **chloral** CCl_3CHO or formaldehyde HCHO, the equilibrium is shifted towards the diol. In its diol form, chloral is known as **chloral hydrate** and serves as a disinfectant. The trade name of formaldehyde is **formalin** which is actually the water solution of formaldehyde.

$$
\underset{\textbf{Chloral}}{\overset{\displaystyle \text{O}}{\underset{|}{CCl_3CH}}} \;+\; H_2O \;\rightleftharpoons\; \underset{\text{OH}}{\overset{\displaystyle \text{OH}}{\underset{|}{CCl_3CH}}}
$$

Chloralhydrate

$$
\underset{\textbf{Formaldehyde}}{\overset{\displaystyle \text{O}}{\underset{|}{HCH}}} \;+\; H_2O \;\rightleftharpoons\; \underset{\text{OH}}{\overset{\displaystyle \text{OH}}{\underset{|}{HCH}}}
$$

Formol (formalin) in water solution

Under the standard conditions, formaldehyde is gaseous. However, its molecules can interact with each other by the mechanism similar to nucleophilic attack. Oxygen of the carbonyl group behaves as the nucleophilic atom. The product of such reaction is polymer, the white powder called **paraformaldehyde**. Hence, formaldehyde can be provided either in the form of the water solution formalin or as paraformaldehyde powder. Pure formaldehyde gas for use in chemical reactions can be prepared by heating paraformaldehyde under vacuum. Similarly, the molecules of acetaldehyde can condense into cyclic trimers giving the substance known as **paraldehyde**.

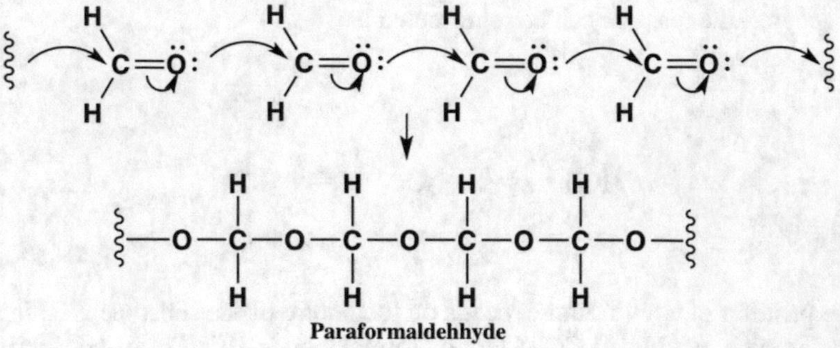

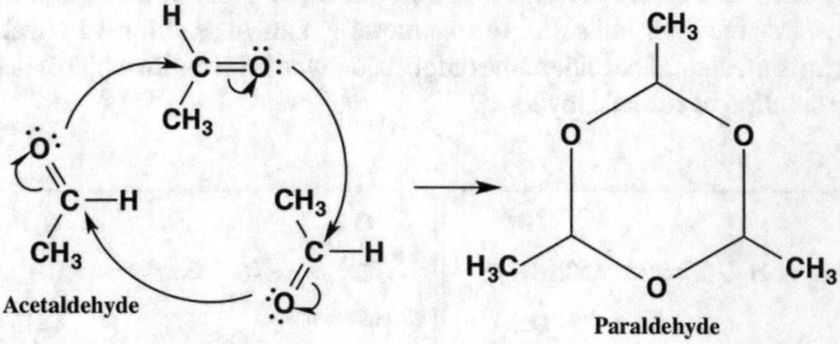

Paraformaldehhyde

Acetaldehyde

Paraldehyde

Alcoholate anions in alcohol solutions react with aldehydes and ketones forming **ketals** and **acetals**, or **hemiacetals** and **hemiketals**. Rates of nucleophilic additions on the carbonyl group are enhanced by protonation of the carbonyl oxygen, so these reactions are **acid-catalyzed**.

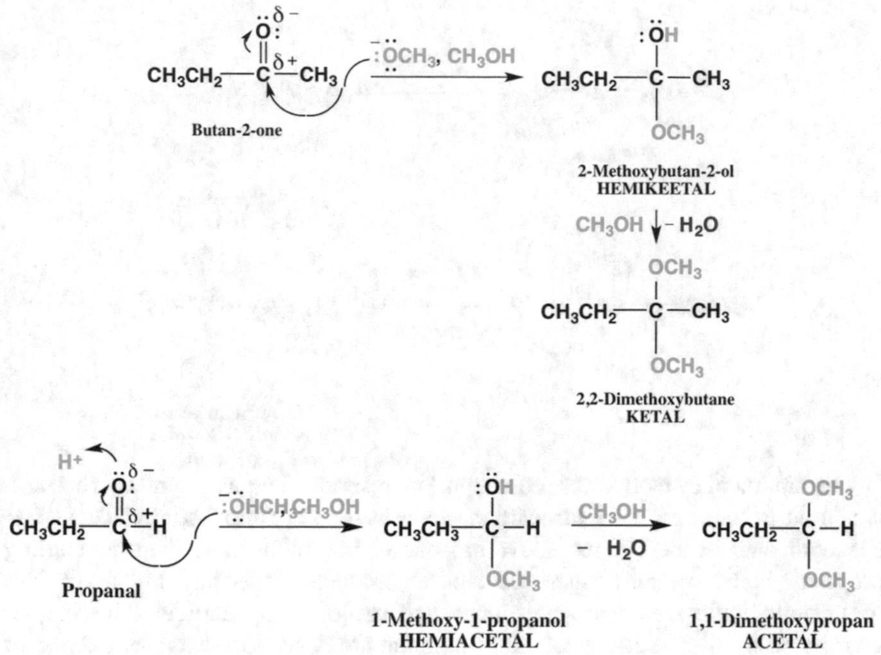

$$CH_3CH_2-\overset{\overset{\overset{\delta-}{\ddot{O}:}}{\|}}{\underset{}{C}}\overset{\delta+}{-}CH_3 \quad \xrightarrow{\ddot{:}\ddot{O}CH_3,\ CH_3OH} \quad CH_3CH_2-\overset{:\ddot{O}H}{\underset{OCH_3}{\overset{|}{C}}}-CH_3$$

Butan-2-one

2-Methoxybutan-2-ol
HEMIKEETAL

$$CH_3OH \downarrow -H_2O$$

$$CH_3CH_2-\overset{OCH_3}{\underset{OCH_3}{\overset{|}{C}}}-CH_3$$

2,2-Dimethoxybutane
KETAL

$$CH_3CH_2-\overset{\overset{\overset{\delta-}{\ddot{O}:}}{\|}}{\underset{}{C}}\overset{\delta+}{-}H \quad \xrightarrow{\ddot{:}\ddot{O}HCH_3CH_3OH} \quad CH_3CH_3-\overset{:\ddot{O}H}{\underset{OCH_3}{\overset{|}{C}}}-H \quad \xrightarrow[-\ H_2O]{CH_3OH} \quad CH_3CH_2-\overset{OCH_3}{\underset{OCH_3}{\overset{|}{C}}}-H$$

H⁺

Propanal

1-Methoxy-1-propanol
HEMIACETAL

1,1-Dimethoxypropan
ACETAL

Acetals and ketals are unstable in acidic solutions and undergo hydrolysis reverting to the original aldehydes and ketones.

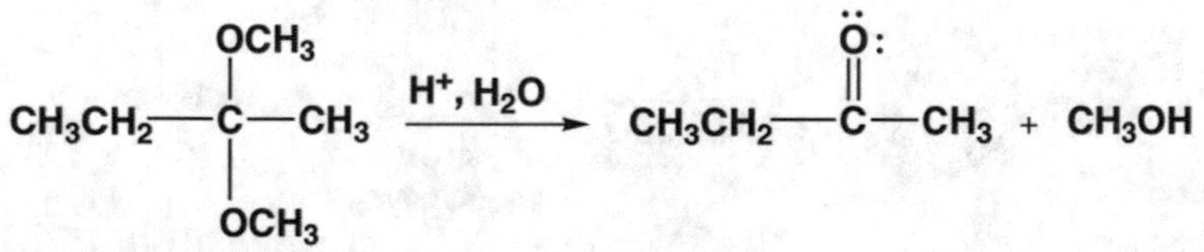

$$CH_3CH_2-\overset{OCH_3}{\underset{OCH_3}{\overset{|}{C}}}-CH_3 \quad \xrightarrow{H^+,\ H_2O} \quad CH_3CH_2-\overset{\overset{\ddot{O}:}{\|}}{C}-CH_3 + CH_3OH$$

The mechanism analogous to acid-catalyzed nucleophilic addition to the carbonyl group applies to the reactions of aldehydes with different nucleophiles like halide or nitrile ions yielding **halohydrins** or **cyanohydrins**.

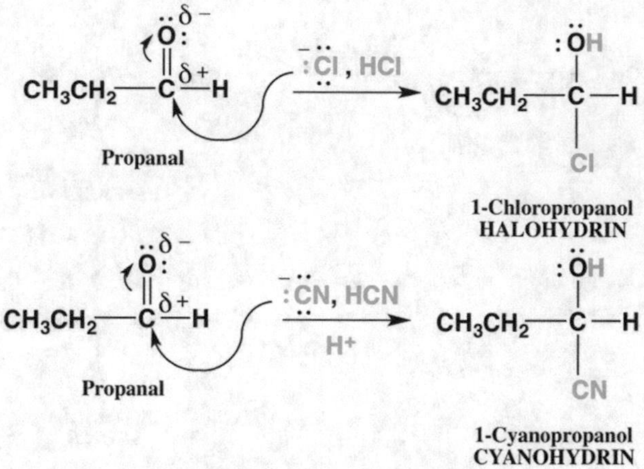

$$CH_3CH_2-\overset{\overset{\textstyle \ddot{O}:^{\delta -}}{\|}}{\underset{}{C^{\delta +}}}-H \quad \xrightarrow{:\ddot{C}l, \ HCl} \quad CH_3CH_2-\overset{:\ddot{O}H}{\underset{Cl}{\overset{|}{C}}}-H$$

Propanal

1-Chloropropanol
HALOHYDRIN

$$CH_3CH_2-\overset{\overset{\textstyle \ddot{O}:^{\delta -}}{\|}}{\underset{}{C^{\delta +}}}-H \quad \xrightarrow[H^+]{:\ddot{C}N, \ HCN} \quad CH_3CH_2-\overset{:\ddot{O}H}{\underset{CN}{\overset{|}{C}}}-H$$

Propanal

1-Cyanopropanol
CYANOHYDRIN

Preparation of cyanohydrins is an important reaction in organic synthesis because it is used to lengthen the hydrocarbon chain by one C-atom: the molecule of 1-cyanopropanol in the scheme above has one carbon atom more than the starting propanal. Some cyanohydrines are natural products, especially the derivatives of **benzaldehyde** which are poisonous and exploited by particular insects for their defense. Toxic activity of these compounds is based on the hydrolysis of cyanohydrine into benzaldehyde and highly toxic HCN.

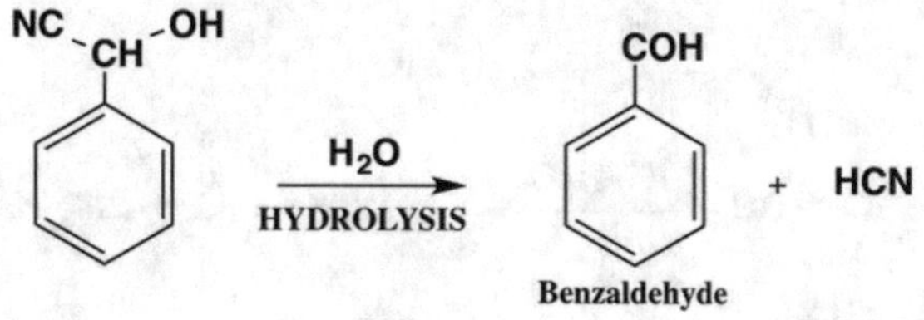

6.1.1   Carbon as a Nucleophile

As we have mentioned at the beginning of this book, the appearance of the large diversity of hydrocarbon compounds is the basic requirement for organic and biological evolution. Aldehydes and ketones undergo simple transformations by which high structural diversity can be obtained starting with only a few simple compounds. In alkaline medium carbonyl compounds can act as acids, i.e., they behave as proton donors (see following scheme). Stable conjugate base can appear only after removing the proton from the C-atom which is directly bound to the carbonyl group.

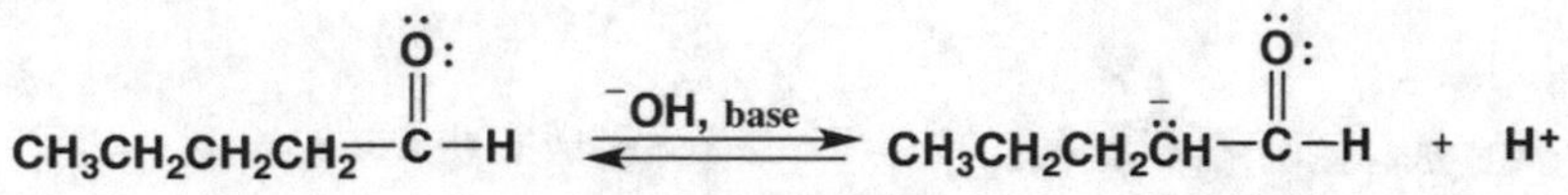

$$CH_3CH_2CH_2CH_2-\overset{\overset{\ddot{O}:}{\|}}{C}-H \quad \xrightarrow{\text{}^-OH, \text{ base}} \quad CH_3CH_2CH_2\overset{-}{C}H-\overset{\overset{\ddot{O}:}{\|}}{C}-H \quad + \quad H^+$$

Pentanal

Stability of the conjugated base arises from the electron delocalization that can be recognized by using the resonance structure method. Acids in which the proton is removed from carbon atom are called **carbon acids** and their conjugate base is called the **enolate ion**.

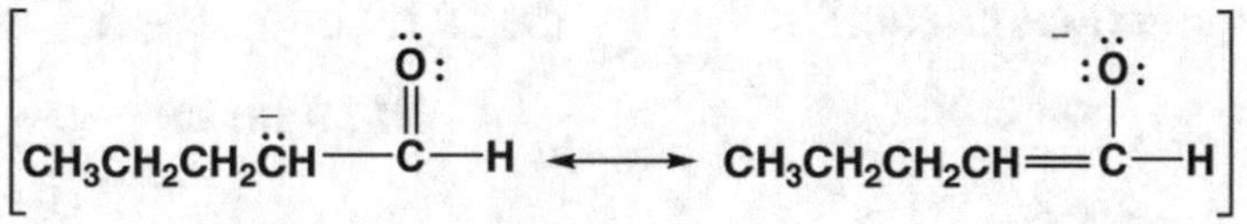

$$\left[CH_3CH_2CH_2\overset{-}{C}H-\overset{\overset{\ddot{O}:}{\|}}{C}-H \quad \longleftrightarrow \quad CH_3CH_2CH_2CH=\overset{\overset{:\ddot{O}:}{}}{C}-H \right]$$

The anions which would be generated by removing proton from the carbon atoms which are not neighbors of the carbonyl group are unstable because their electron pair is localized. No resonance structures can be written in such cases (see the structure below).

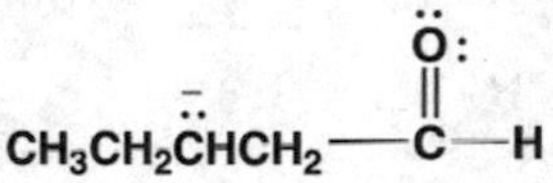

$$CH_3CH_2\overset{..}{\overset{-}{C}}HCH_2-\overset{\overset{\ddot{O}:}{\|}}{C}-H$$

Since under the basic conditions the aldehyde molecules partially dissociate, their mixture contains two molecular species, dissociated and undissociated. The dissociated molecule, i.e., the enolate ion, can behave as nucleophile because of the electron pair localized on its carbon atom. This nucleophile can attack the carbonyl group of the undissociated aldehyde yielding the product with two functional groups, aldehyde and alcohol. Such compounds are named **aldols** and the corresponding reaction is called **aldol condensation**.

$$CH_3CH_2CH_2CH_2—\overset{\overset{\displaystyle ..O:}{\|}}{C}—H$$

$$CH_3CH_2CH_2\overset{..}{C}H———\overset{\overset{\displaystyle }{}}{C}—H$$
$$\overset{\displaystyle }{\underset{\displaystyle O:}{}}$$

$$CH_3CH_2CH_2CH_2—\overset{\overset{\displaystyle :\overset{..}{O}:{}^-}{|}}{C}—H$$

$$CH_3CH_2CH_2CH———C—H$$
$$\overset{\displaystyle }{\underset{\displaystyle O:}{\|}}$$

$+ H^+$

$$CH_3CH_2CH_2CH_2—\overset{\overset{\displaystyle :\overset{..}{O}H}{|}}{C}—H$$

$$CH_3CH_2CH_2CH———C—H$$
$$\overset{\displaystyle }{\underset{\displaystyle O:}{\|}}$$

$$CH_3CH_2CH_2CH_2\overset{\overset{\displaystyle OH}{|}}{C}H\overset{\overset{\displaystyle O}{\|}}{C}H$$
$$\underset{\displaystyle CH_3CH_2CH_2}{|}$$

ALDOL

After removing the molecule of water, the aldol is transformed into the corresponding unsaturated aldehyde.

$$CH_3CH_2CH_2CH_2\overset{\overset{\displaystyle OH}{|}}{C}H\overset{\overset{\displaystyle O}{\|}}{C}H \quad \underset{- H_2O}{\longrightarrow} \quad CH_3CH_2CH_2CH_2CH=\overset{\overset{\displaystyle O}{\|}}{C}CH$$
$$\underset{\displaystyle CH_3CH_2CH_2}{|} \qquad\qquad\qquad \underset{\displaystyle CH_3CH_2CH_2}{|}$$

ALDOL

The example given is the reaction in which the molecule with five C-atoms in the hydrocarbon chain (pentanal) gets converted to the molecule containing nine atoms in the longest chain. Let us imagine a mixture of two different aldehydes, A and B. Since these aldehydes can form two enolate nucleophiles and two undissociated molecules, the cross reactions must yield four products: A + A, B + B, A + B and B + A. The number of products of aldol condensation follows the n^2 rule where n is the number of different aldehydes used. For instance, the mixture of ten different aldehydes will produce a hundred different products. If unsymmetrical ketones with two different carbons next to the carbonyl group are added to the starting mixture the diversity of products in the mixture is further enhanced. This example demonstrates how large number of diverse structures can originate from a small number of starting molecules.

Molecules with carbon atom as a nucleophilic center also appear in the reactions of carbonyl compounds with organometallic compounds. Such reactions are very useful in organic synthesis. The best-known reagents with such carbon nucleophiles are alkyl- or arylmagnesium halides, which in the honor of **Viktor Grignard** are called **Grignard reagents**. The reagents can be easily prepared by direct reactions of alkyl- or arylhalide with magnesium metal. Since Grignard reagents are, like many organometallic compounds, unstable and sensitive to moisture the reaction must be performed in dry diethyl-ether.

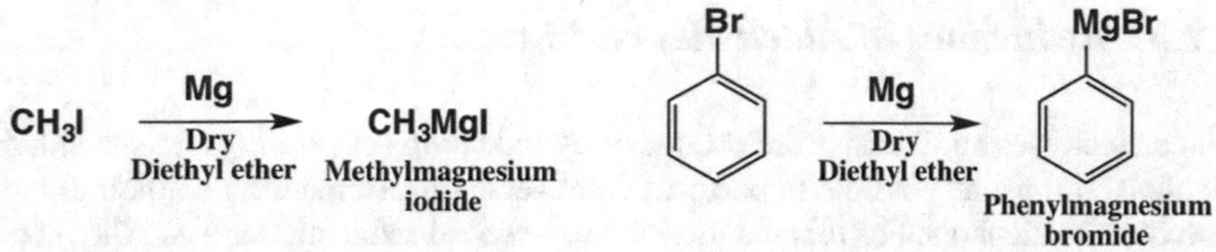

The carbon-magnesium bond is polar with partial negative charge on the carbon atom. Consequently, this nucleophilic C-atom attacks the carbon atom of the carbonyl group while the positive magnesium halide residue bonds to oxygen. The intermediate produced in the next reaction step is hydrolyzed to the corresponding alcohol. It is important to observe that the product molecule has one additional methyl group.

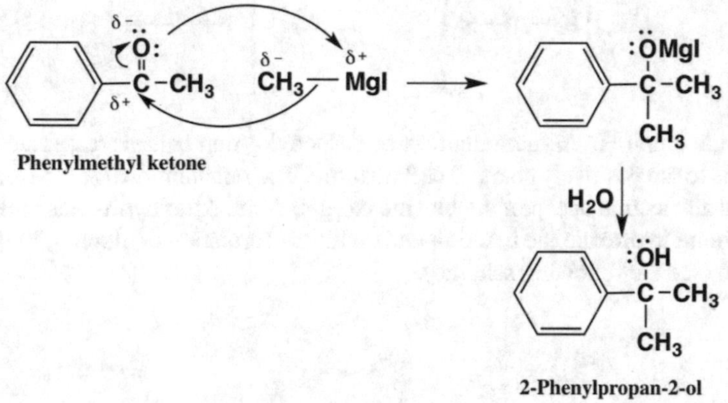

Phenylmethyl ketone

2-Phenylpropan-2-ol

Starting with alkyl halides we can prepare the appropriate Grignard reagents and in combination with different aldehydes or ketones, synthesize a series of compounds with branched hydrocarbon chains.

6.1.2 Condensations with Amines

Aldehydes and ketones react readily with equimolar amounts of primary amines yielding imines and water. **Imines** (also called **Schiff bases**) can be considered to be amines with carbon–nitrogen double bonds.

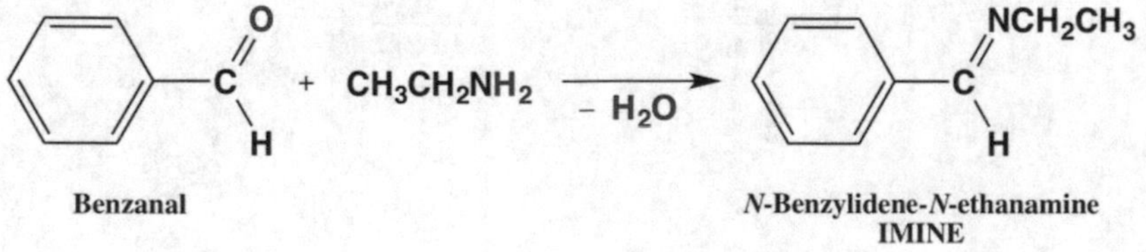

Benzanal

N-Benzylidene-N-ethanamine
IMINE

6.1.3   Reductions of Aldehydes and Ketones

Since aldehydes and ketones are produced by oxidations of primary and secondary alcohols, it must be possible to perform the reverse transformation i.e., their reduction back to alcohols. For this reaction we use special reducing reagents, the metal hydrides because the reduction in organic reactions involves addition of hydride ion H^-. The most important reducing agents are **lithium aluminium hydride**, $LiAlH_4$, **sodium boron hydride,** $NaBH_4$ and **sodium hydride**, NaH. Molecules of $LiAlH_4$ and $NaBH_4$ consist of lithium or sodium cations and AlH_4^- or BH_4^- anions.

$$Li^+ \begin{bmatrix} H \\ | \\ H-Al-H \\ | \\ H \end{bmatrix}^- \qquad Na^+ \begin{bmatrix} H \\ | \\ H-B-H \\ | \\ H \end{bmatrix}^-$$

Hydride anion H^- in the reduction of carbonyl group behaves as a nucleophile and binds to the positively charged carbon atom. The remaining AlH_3 is a Lewis acid which as an electron acceptor attacks the oxygen atom. After hydrolysis and release of aluminum hydroxide the reaction ends with the formation of alcohol, in this case propanol (see the following scheme).

Reduction of ketone follows the same mechanism. In the following example acetone is reduced to propane-2-ol.

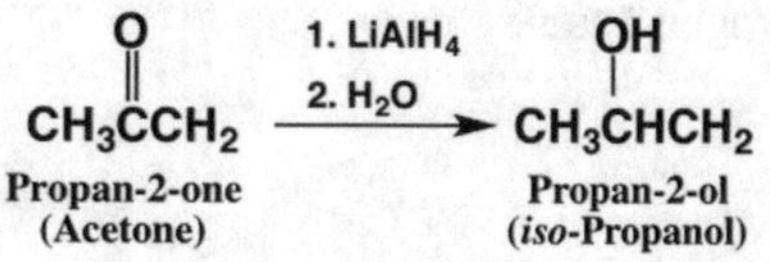

6.2 Carboxylic Acids

In organic chemistry, the oxidation process can also imply addition of oxygen atoms. In such a way the aldehyde group is oxidized to the **carboxyl group** by adding an oxygen atom. The compounds with carboxyl group belong to **carboxylic acids**. Oxidative agents for this transformation include sulfuric acid, solutions of potassium permanganate $KMnO_4$, potassium bichromate $K_2Cr_2O_7$ or chromium trioxide CrO_3. The following example represents the oxidation of pentanal to the corresponding carboxylic acid.

Pentanal $KMnO_4, H_2SO_4$ / $H_2O, 20°C$ Pentanoic acid

Carboxyl group

Justus von Liebig and **Bernhard Tollens** have discovered that milder agents such as silver oxide, Ag_2O oxidize aldehydes to carboxylic acids. This reaction, which is characteristic for aldehydes serves as a specific qualitative test for this class of compounds and is known as **Tollens test**. Metallic silver being one of the byproducts precipitates and forms characteristic silver mirror that can be easily observed.

Benzanal (Benzaldehyde)   1. Ag_2O, NaOH, H_2O 2. H_3O^+ Benzoic acid + **2Ag (silver mirror)**

Alternative method for the preparation of carboxylic acids is the reaction of nitriles with sulfuric acid. This reaction is important in organic synthesis because it can be used to extend the hydrocarbon chain by one carbon atom. Nitriles can be obtained from the corresponding alkyl halides in the reaction with NaCN by nucleophilic substitution. Nitrile can easily be oxidized to carboxylic acid.

$CH_3CH_2CHCH_2Br$ | CH_3 NaCN / – Br⁻ $CH_3CH_2CHCH_2CN$ | CH_3 H_2O, H_2SO_4 $CH_3CH_2CHCH_2C$ | CH_3 $\diagup O \diagdown OH$

2-Methylbutanebromide   3-Methylpentanenitrile   3-Methylpentanoic acid

Many essential natural products that are ubiquitous in living organisms are carboxylic acids. The simplest (parent) carboxylic acid is **formic acid** (*acidum formicum*) which is named also **methanoic acid** in the IUPAC nomenclature. This acid is the main component of the poison of ants.

Acetic acid (*acidum aceticum*) with IUPAC name **ethanoic acid** is used as everyday food additive. **Citric acid**, which is also a natural product is the polyfunctional compound with three carboxyl and one OH group. Molecule of simple **oxalic acid** consists of only two carboxyl groups. The oxalic acid salts are frequently present in kidney stones.

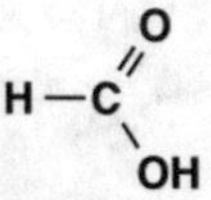

| Methanoic acid (Formic acid) | Ethanoic acid (Acetic acid) | Oxalic acid | Citric acid |

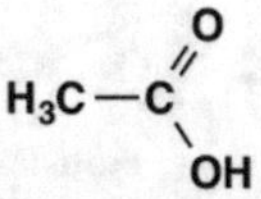

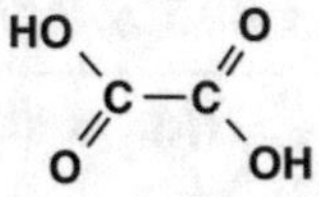

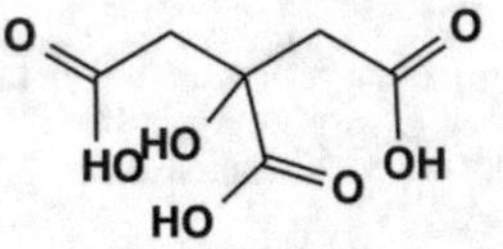

Many carboxylic acids such as **cholic acid** or long-chained **fatty acids** are complex molecules. In contrast to the saturated fatty acids, molecules of unsaturated fatty acids always have carbon–carbon double bonds. Interestingly, the molecules of fatty acids, without exception, have even number of C-atoms. This is an indication that the natural pathways for biosynthesis of these classes of compounds are the same. The role of fatty acids and other natural products in living organisms will be discussed later in this book.

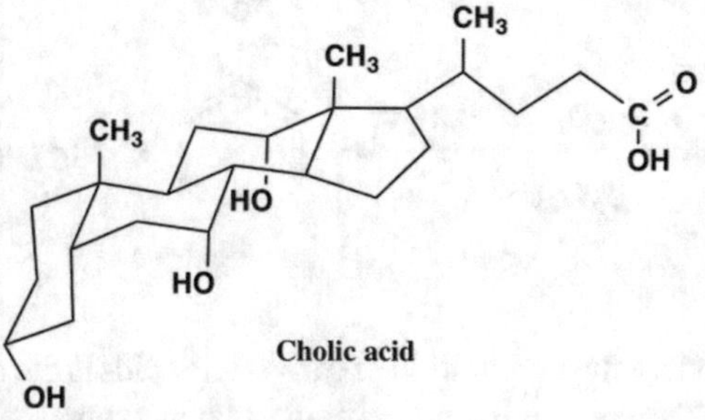

Cholic acid

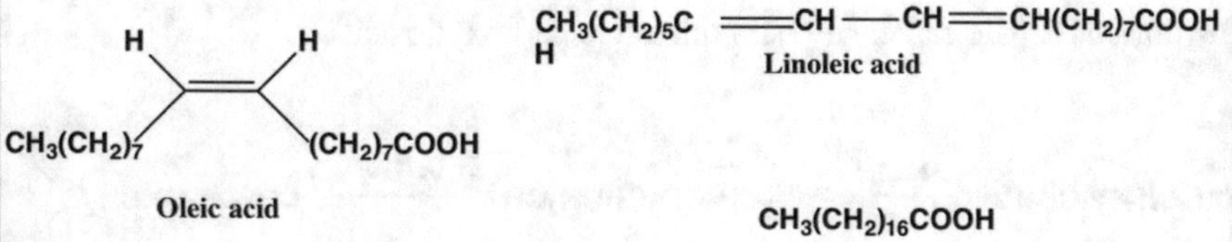

Oleic acid

Linoleic acid

Stearic acid

The rules for naming carboxylic acids include principles of organic and principles of inorganic nomenclature. The name of the acid consists of the name of the longest hydrocarbon chain and the prefix **–oic acid**. In numbering C-atoms of the hydrocarbon chain, the smallest number must be assigned to the carboxylic group. If carboxylic group is a substituent than its carbon atom does not belong to the hydrocarbon chain. Instead, the name is formed with the prefix **carboxylic acid** (as cyclohexanecarboxylic acid in the following example).

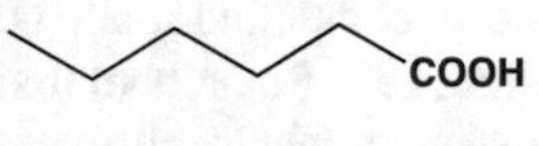

Hexanoic acid 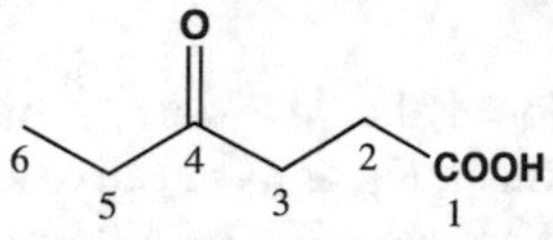   **4-Oxohexanoic acid** 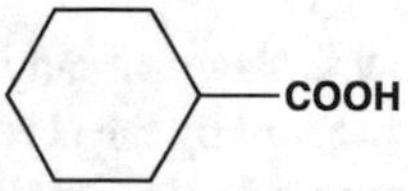   **Cyclohexanecarboxylic acid**

Carboxylic acids are substances with high boiling points: formic acid 101 °C, acetic acid 118 °C and propanoic acid 141 °C. This is the consequence of dimerization and formation of molecular aggregates via hydrogen bonds. Even in the gas phase, the molecules can remain in dimeric form.

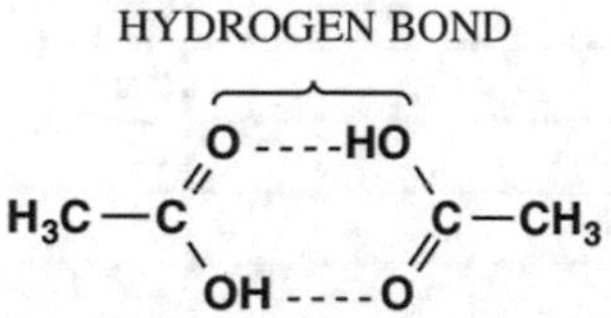

HYDROGEN BOND

**Dimer of the acetic acid molecules
interconnected by hydrogen bonds**

Acidic behavior of carboxylic acids as good proton donors is the consequence of stabilization of the corresponding conjugated base (**carboxylate anion**). As can be seen from the resonance structures presented below, the electrons in the carboxylate anion are delocalized.

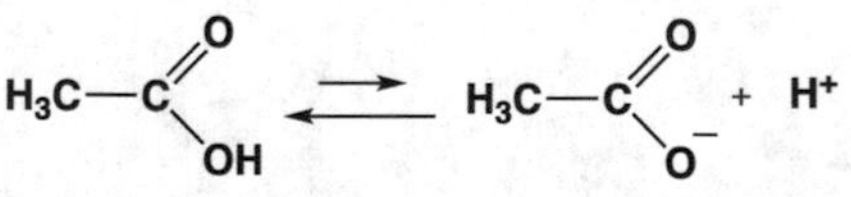

Ethanoic acid   **Ethanoate anion** 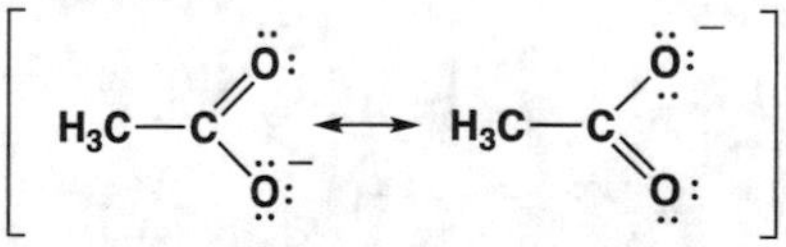   **Resonance structures of ethanoate anion**

The structures show that negative charge is equally distributed on both oxygen atoms. The charge distribution in the molecule can be calculated by using quantum–mechanical methods and molecular modeling. When the carboxylate anion (in our case ethanoate anion) is modeled by such methods we obtain the diagram shown below with regions of highest electron density (negative charge) are shown in red:

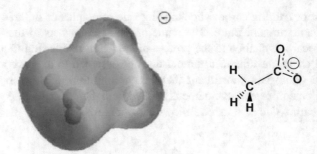

We have already mentioned that the acidity of organic compounds can be expressed by the acidity constant, K_a, or its negative logarithm, pK_a. Remember that stronger acids have lower pK_a values. The relationship between molecular structure and acidity for different carboxylic acids, which is shown in the table below can be used to estimate the structure–activity relationships of organic compounds in general.

Carboxylic acid	pK_a
CH_3COOH	4.7
CH_3CH_2COOH	4.9
CH_2FCOOH	2.6
$CH_2ClCOOH$	2.9
$CHCl_2COOH$	1.3
CCl_3COOH	0.9

The first pattern that is evident from these data is the influence of halogens on acidity. As halogens become more electronegative or when their number increases, the carboxylic acid becomes stronger (it has lower pK_a). Since it is not possible to write resonance structures which would include electrons from the halogens, the electron delocalization in such anions must have a different origin. To get a better insight into this alternative mode of charge delocalization, let us discuss additional examples.

pK_a 2,89   4,05   4,53

Acidity amongst the three chlorobutanoic acids decreases as the chlorine atom moves away from the carboxyl group. The weakest is 4-chlorobutanoic acid with pK_a 4.53. It is known that since the electronegative halogens attract electron density, their neighboring atoms become partially positively charged. Such effect is most pronounced on the first neighboring C-atom and it gets smaller along the hydrocarbon chain, as it is shown in the scheme below.

$$\text{Cl} \underset{\overset{\delta-}{\longleftarrow}}{\underline{\hspace{1cm}}} \text{C} \underset{\overset{\delta+}{\longleftarrow}}{\underline{\hspace{1cm}}} \text{C} \underset{\overset{\delta\delta+}{\longleftarrow}}{\underline{\hspace{1cm}}} \text{C} \underset{\overset{\delta\delta\delta+}{\longleftarrow}}{\underline{\hspace{1cm}}} \text{C} \underline{\hspace{1cm}} \overset{\delta\delta\delta\delta+}{}$$

Charge transfer through σ-bond

INDUCTIVE EFFEKT

Negative charge that is concentrated on the Cl-atom induces progressively weaker positive charges on the rest of the chain. Such electron delocalization is achieved through σ-bonds and is called **inductive effect**. Since the chlorine atom in the series of chlorobutanoic acid anions is located at different distances from the carboxylic group, its inductive stabilizing influence on the anion is also different: the strongest is in 2-chlorobutanoic acid with the pK_a value 2.89.

In summary, the carboxylic acid is stronger if after the removal of the proton, it can form a more stable conjugated base, the carboxylate anion. The anion is stabilized by electron delocalization over the largest possible region of the molecule. The **charge delocalization** can occur in two ways: through the π-electron system by **resonance effect** or through the σ-electron network by **inductive effect**. The resonance effect has already been explained in the section on alcohols when we discussed the acidity of phenol (vide infra).

In the analogy with inorganic acids, the carboxylic acids can be neutralized with bases. The names of salts which are formed after neutralization follow inorganic nomenclature.

$$H_3C - C \overset{O}{\underset{OH}{\big\backslash}} \quad + \quad NaOH \quad \longrightarrow \quad H_3C - C \overset{O}{\underset{O^- Na^+}{\big\backslash}} \quad + \quad H_2O$$

Ethanoic acid　　　　　　　　　　　　　**Sodium ethanoate**
(Acetic acid)　　　　　　　　　　　　　**(Sodium acetate)**

$$H - C \overset{O}{\underset{OH}{\big\backslash}} \quad + \quad KOH \quad \longrightarrow \quad H - C \overset{O}{\underset{O^- K^+}{\big\backslash}} \quad + \quad H_2O$$

Methanoic acid　　　　　　　　　　　　**Potassium methanoate**
(Formic acid)　　　　　　　　　　　　　**(Potassium formate)**

Of special interest are sodium and potassium salts of the long-chained fatty acids, the compounds known under the traditional name of **soaps** (see the next scheme). The chemical behavior of these molecules exhibits dual nature. The long hydrocarbon chain is typical of the organic component which is insoluble in water. Because hydrocarbon chains repel water molecules, they are called **hydrophobic**. On the other side of the molecule is the carboxylate group which is with the metal ion bound by ionic bond, and such compounds are readily soluble in water. They attract water molecules and we call them **hydrophilic**. The molecules of soaps possess a hydrophilic head and a hydrophobic tail.

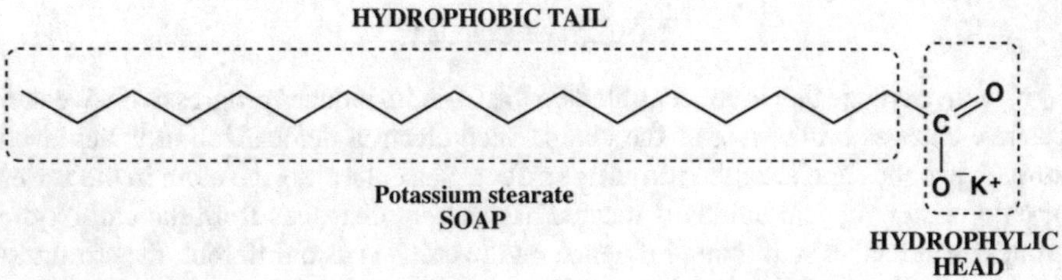

HYDROPHOBIC TAIL

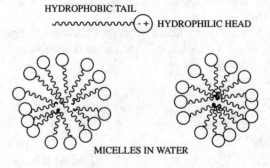

**Potassium stearate
SOAP**

**HYDROPHYLIC
HEAD**

Because of their specific structures the soap molecules exhibit special behavior in water solutions, they undergo **self-organization**. The molecules form complex structures called **micelles** in which the hydrophobic tails are oriented towards the middle of the particle and the hydrophilic heads are pointed towards the water solvent.

HYDROPHOBIC TAIL
∿∿∿∿∿∿(- +) HYDROPHILIC HEAD

MICELLES IN WATER

The medium inside micelles is hydrophobic and behaves as a non-polar organic solvent. Because of their dual nature the soaps are used for removing fatty impurities, which cannot be washed away with water. Micelle absorbs the impurity into its hydrophobic central part, but the whole particle is hydrophilic and can be easily washed away by water. Similar behavior is shown by detergents, the compounds that differ from soaps by the nature of their hydrophilic head. The most popular detergents are salts of organic sulfates or quaternary ammonium salts (see for example **SDS**, and **CTAB** in the following scheme). Hydrophilic head in SDS is an anion so these **micelles are called anionic**, while the head in CTAB is cationic and these are named **cationic micelles**. Substances whose molecules can form such micellar structures belong to the large class of compounds, **surfactants**, the substances of great practical and industrial significance.

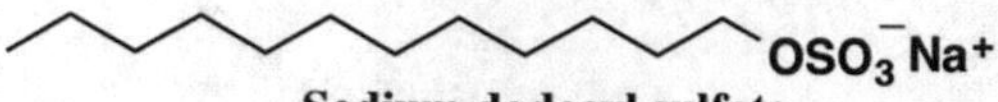

Sodium dodecyl sulfate (SDS) — $OSO_3^- Na^+$

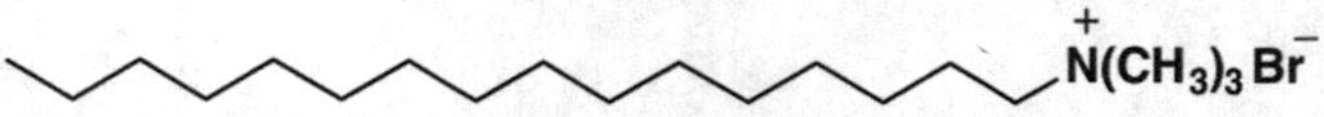

Cetyltrimethylammonium bromide (CTAB) — $N(CH_3)_3^+ Br^-$

Chapter 7
Stereochemistry, Symmetry and Molecular Chirality

Joseph Achille LeBel was one of the most important pupils and collaborators of **Louis Pasteur**. Both of them were interested in the salts of tartaric acid (contained in wine) the tartarates. During the isolation of ammonium-sodium-tartarate, Pasteur and LeBel have observed that the compound crystalizes in two kinds of crystals (see figure below). Although these crystals have equal shapes, they differ in such a way as to be mirror picture of one another.

H. Vančik, *Basic Organic Chemistry for the Life Sciences*,
https://doi.org/10.1007/978-3-030-92438-6_7

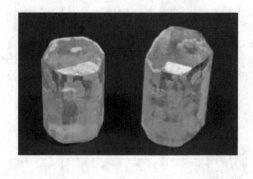

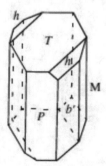

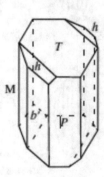

This is similar to the spatial relationship between the left and the right hand which cannot be superimposed by any possible reorientation in space. Since LeBel's crystals reveal the same property they have been called **chiral** i.e., similar to the hand (Greek word for hand is $\chi\varepsilon\acute{\iota}\rho$). Almost at the same time, **Jean Baptist Biot**, the physicist who also collaborated with Pasteur, discovered the polarized light. Today we know that light is an electromagnetic wave described as a vibration of two perpendicular vectors, electric and magnetic. If both vectors vibrate in fixed planes, the light is **linearly or plane polarized**. Natural visible light is a mixture of infinite linearly polarized waves. Some substances like polymers and certain crystals transmit only one plane of the propagating light. They are called **polarizers**.

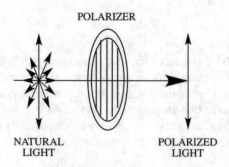

By passing through the polarizer, the nonpolarized light becomes linearly polarized. If the polarized light passes through two polarizers in sequence, the extent of passage will depend on the relative orientation of the two polarizers. When the polarizing planes in both polarizers are parallel the light will pass through both of them, but if the planes are perpendicular the light cannot pass through the second polarizer.

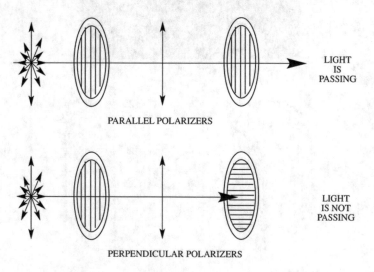

LeBel has discovered that if the solution of only one of the two types of crystals (for instance one at the left side of the diagram) is placed between the two polarizers the polarized light passes through both polarizers. However, for this to occur, the polarizers are not oriented in parallel, but are rotated with respect to each other by a certain angle. The solution of the crystals at the right-hand side of the diagram rotates the polarized light by the same angle but in the opposite direction. Such rotation of the plane of polarization is known as **optical activity**. Since the solutions behave as "left" and "right" in the same way as their crystals are left- and right-handed, the optical activity must have its origin in the structure of molecules. LeBel concluded that molecules must also be left- and right-handed, i.e., they must be chiral. Accordingly, the molecules must have spatial structures which are mirror images of one another. Taking into account this evidence, LeBel has discovered that organic molecules may be chiral only if the carbon atom is located at the center of tetrahedron. This discovery gave birth to the field of **stereochemistry**. The same idea about tetrahedral structure and chirality of organic molecules has also been proposed by van't Hoff, who developed his theory independently and using quite different approach.

Besides tetrahedral structure, the chirality of molecule requires one additional condition: all four substituents bound to the central carbon atom must be different. Such carbon atom with four different substituents induces chirality of the entire molecule and it is called **stereogenic**, **chiral** or **asymmetric center**. If two or more substituents are identical the molecule and its mirror image can be superimposed and the molecule is not chiral. Some chiral and non-chiral molecules are represented bellow.

Let us examine why only molecules with four different substituents are chiral? What is the fundamental geometric property which the molecule, crystal or hand must possess in order to be chiral? These phenomena can be examined within the **theory of symmetry**. Symmetry is the property which ensures that the figure or geometrical body remains unchanged under particular spatial operation which is called a **symmetry element**. We say that the object is symmetric in relation to these symmetry elements. Some of the symmetry elements were already described in the chapter on the symmetry of molecular orbitals. Here, let us introduce additional symmetry elements:

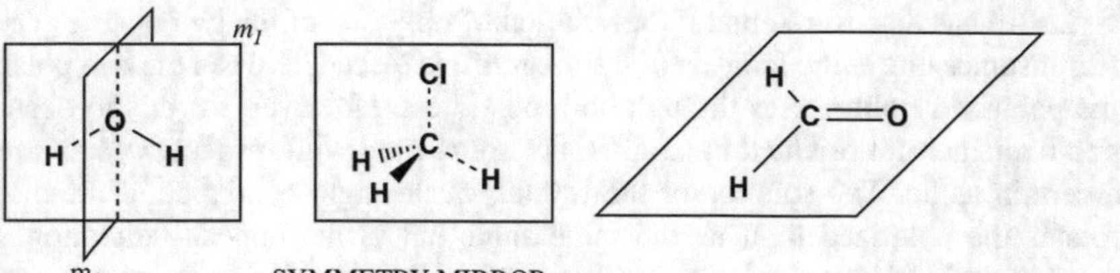

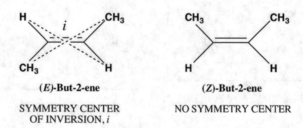

SYMMETRY MIRROR, m

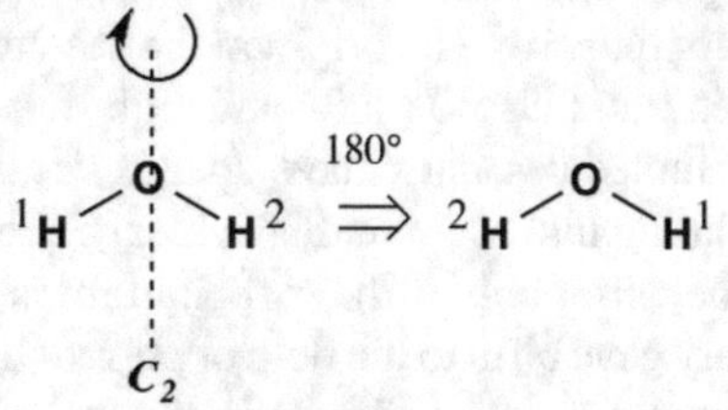

AXIS of SYMMETRY, C_2

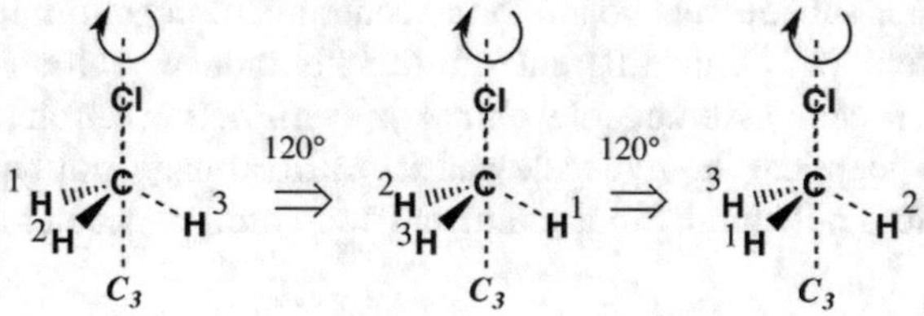

AXIS of SYMMETRY, C_3

(E)-But-2-ene   **(Z)-But-2-ene**

SYMMETRY CENTER   NO SYMMETRY CENTER
OF INVERSION, i

The scheme above represents three symmetry elements: **symmetry plane**, **symmetry axis** and the **center of symmetry**. We can recognize that the water molecule, methyl-chloride and formaldehyde all have symmetry planes. The water molecule and formaldehyde also possess one symmetry axis that leaves the molecule unchanged when rotated by 180°. The symmetry axis around which the object can be rotated by 180° angle and remain unchanged is called **second order symmetry axis** or **digyre**. The methyl chloride molecule has the symmetry axis with 120° rotation. It is called **third order axis** or **trigyre**. In the bottom row of the diagram above (E)-but-2-ene has the **center of symmetry,** but (Z)-but-2-ene does not. Centre of symmetry is the point (labeled with the symbol *i*) through which each atom can be reflected into the equivalent atom with both atoms being at the same distance from *i*. In this way the H-atom in (*E*)-but-2-ene is reflected through *i* into another H-atom. The same principle is valid for methyl groups. The chiral molecules have special symmetry properties because they do not have any symmetry planes as for instance the molecule CHClBrF. Clearly, all molecules in which the carbon atom is bound with the double bond are not chiral and such carbon is not a center of chirality. We know that each C-atom of the double bond is coplanar with three of its neighbors and this plane is at the same time the plane of symmetry. The molecule of formaldehyde is a good example. The four different substituents bound to the chiral carbon need not be the atoms of four different chemical elements. Chirality is present if substituents are different groups or if they have different constitutions. Every chiral molecule with one stereogenic center can have two isomers called **enantiomers**. Enantiomers are related to each other as the object is to its mirror image. Since enantiomers are isomers, they can be transformed into each other only by a chemical reaction i.e., by the breaking and forming of chemical bond. Transformation of one enantiomer into another can involve breaking of any two chemical bonds to the stereogenic centre, switching the positions of corresponding substituents and the forming of two new chemical bonds.

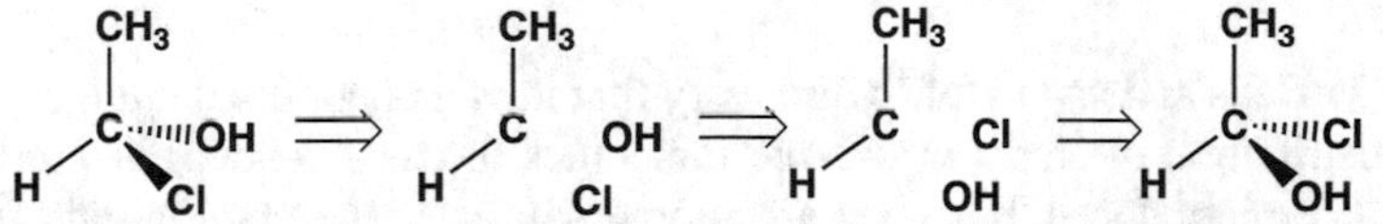

| ONE ENANTIOMER | REPLACEMENTS OF SUBSTITUENTS (CHEMICAL REACTION) | ANOTHER ENANTIOMER |

If the molecule has more than one stereogenic center, every center can exist in two enantiomeric forms. For *n* stereogenic centers, the number of stereoisomers is 2^n. Stereoisomers of molecules with more than one asymmetric carbon atom are called **diastereomers**. In constitutional molecular formulas the stereogenic centers are usually labeled with asterisks. Some examples of molecules having diastereomers are shown below.

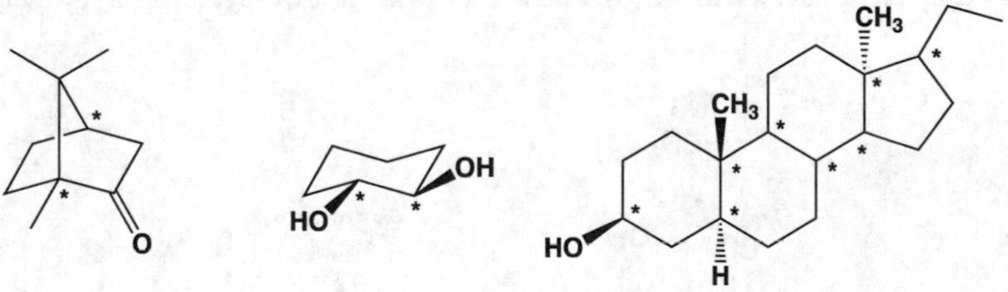

In search for chiral C-atoms, it is convenient first to eliminate carbon atoms which cannot be chiral centers. These are for instance CH_2, or CH_3 groups as well as all the C-atoms connected with double or triple bonds.

As it has been already mentioned, the enantiomers rotate the plane of polarized light by a particular angle. Usually, the label (+) is used for the enantiomer whose solution rotates the plane to the right and the label (−) for the one which rotates it to the left. The mixture in which both enantiomers are in equal concentrations does not exhibit optical rotation. Such solution that is known as **racemate** does not show rotation of the plane polarized light. This is because as one enantiomer rotates to the right by a given angle, the other enantiomer rotates to the left by the same angle so that the rotations cancel each other. Optical rotation is the characteristic property of a particular compound and is called **specific rotation**, $[\alpha]_D{}^{25}$. Its value is calculated from the measured optical rotation angle α, the length of the measuring container l in dm and the concentration of the sample c in g/100 cm^3. The index D refers to the type of light used in the measurement. In the instruments used for measuring optical rotation called polarimeters, the source of light is sodium lamp which emits the specific light called the D-line. Additionally, the superscript 25 indicates that the measurement is made at the 25 °C.

$$[\alpha]_D^{25} = \frac{100[\alpha]}{cl}$$

After LeBel's and van't Hoff's discovery that the tetrahedral structure underpins spatial distribution of atoms in organic molecules the new concept of **configuration** has been introduced. Since the enantiomers differ in their configuration it was necessary to develop the system of nomenclature by which their configurations can be labeled unequivocally. Cahn, Ingold and Prelog have invented the **system of absolute configuration**, the method similar to the procedure which we have already discussed in the previous chapters. To determine the absolute configuration of the molecule it is necessary to follow certain procedure which is similar to the determination of Z and E isomers of alkenes.

1. Determination of the **Priorities**.

The highest priority pertains to the atom with the highest atomic number. In the examples given below, chlorine with the highest atomic number has the priority (1) followed by oxygen (2), carbon (3), and the atom with lowest priority hydrogen (4).

2. Molecular orientation

The molecule must be oriented such a way that the chiral carbon atom is in front and the substituent with the lowest priority (in our example hydrogen) behind this carbon. Frontal atoms and groups are labeled in blue in the next scheme.

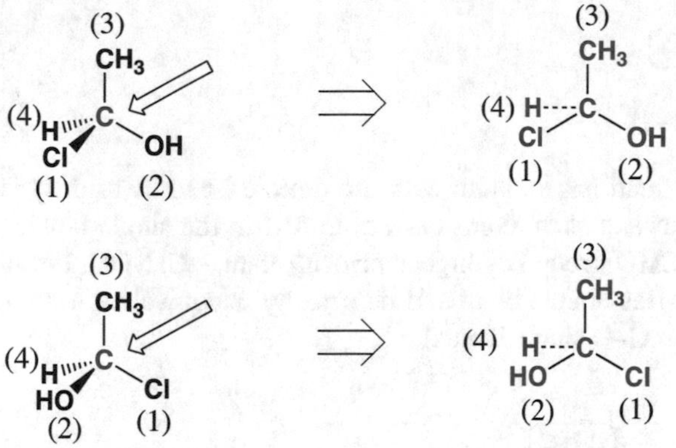

3. Determination of the direction of rotation

The molecule is rotated around the axis between the chiral center and the substituent of the lowest priority (see diagram below).The direction of rotation follows the order of priorities of the three substituents. If the rotation has clockwise orientation the molecule has **(R) absolute configuration** and if the rotation is anticlockwise the molecule will have **absolute configuration (S)**.

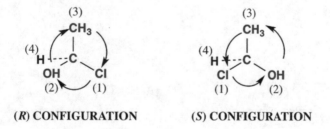

(R) CONFIGURATION   **(S) CONFIGURATION**

4. In most organic molecules the atoms directly bound to the chiral carbon are identical. In that case, the rule must be extended so that we look at the atomic number of the next atom along the chain. In the following example we distinguish between the C-atoms in groups CH_3 and CH_3CH_2. In the CH_3 group, the C-atom bound to the stereogenic center is also bound to three hydrogen atoms. In the CH_3CH_2 group the corresponding atom is bound to the atoms H, H, C. Therefore, since in the CH_3CH_2 group the next atom along the chain is C rather than H, the CH_3CH_2 group has higher priority than CH_3. Hence, the configuration must be (R).

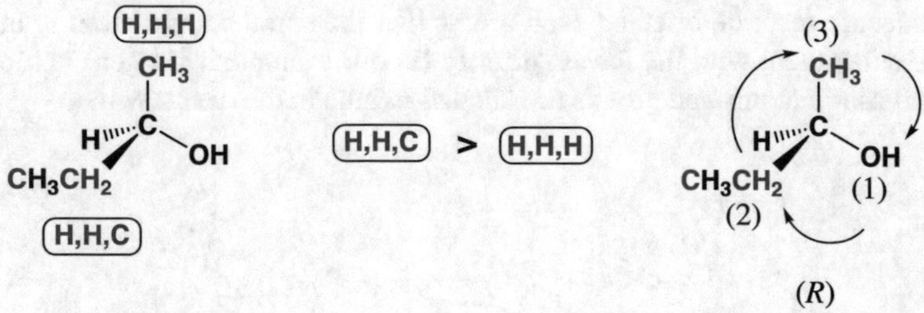

(H,H,H)
CH₃

H''''ᴄ—OH
CH₃CH₂

(H,H,C)

(H,H,C) > (H,H,H)

(3)
CH₃

H''''ᴄ—OH
CH₃CH₂ (1)
(2)

(R)

5. If the substituent has an atom with the double bond its priority is considered to be equivalent to two such atoms each bound with the single bond. In the example bellow, the −CHO group has higher priority than −CH₂OH, because the oxygen in − CHO carries double bond and its priority is equivalent to two oxygen atoms (counted as two C–O single bonds).

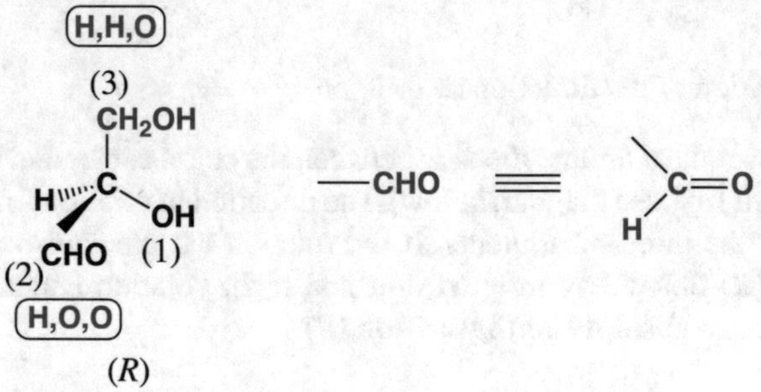

(H,H,O)
(3)
CH₂OH

H''''ᴄ—OH
CHO (1)
(2)

(H,O,O)

(R)

$$—CHO \equiv \underset{H}{\overset{}{>}}C=O$$

Chapter 8
Derivatives of Carboxylic Acids

If the −OH is removed from the carboxylic group, the remaining group has the form
RCO− which is called the **acyl group**. By connecting different atoms or groups
to the acyl group we can obtain various structural derivatives of carboxylic acids,
anhydrides, **esters**, **acyl halides** or **amides**. The substituent R in the following
scheme can be carboxylate RCOO−, alkoxide −OR, halide, for instance −Cl, and
amine −NH_2, NHR or NR_2.

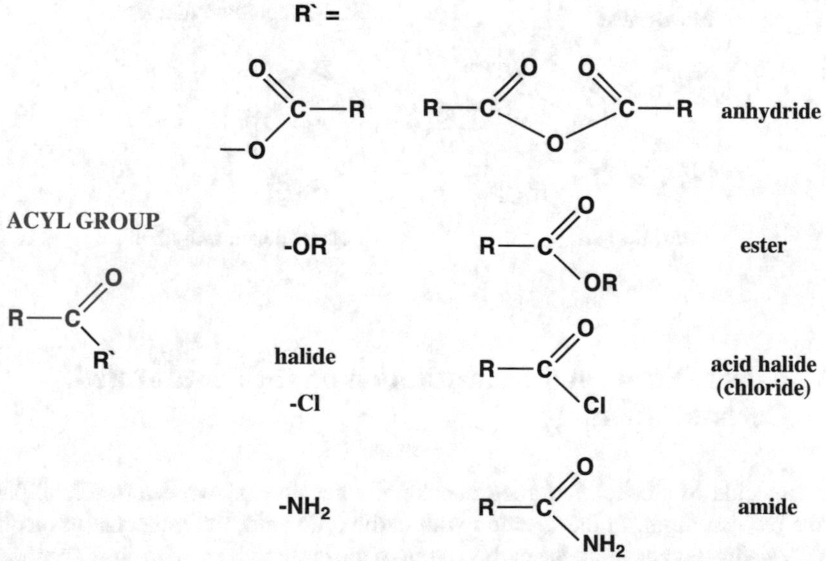

© The Author(s), under exclusive license to Springer Nature Switzerland AG 2022   117
H. Vančik, *Basic Organic Chemistry for the Life Sciences*,
https://doi.org/10.1007/978-3-030-92438-6_8

8.1   Anhydrides

Removing water molecule from the acid yields an anhydride. Inorganic anhydrides are oxides, so SO_3 is the anhydride of sulfuric acid H_2SO_4. Upon removing water from the carboxylic acids, the molecules condense yielding anhydride. If the molecule has only one carboxyl group the condensation involves two molecules, as in the case of acetic acid where two molecules form **acetanhydride**. Acids with two carboxylic groups, such as **maleic acid** or **phthalic acid** form corresponding cyclic structures after removing water molecule.

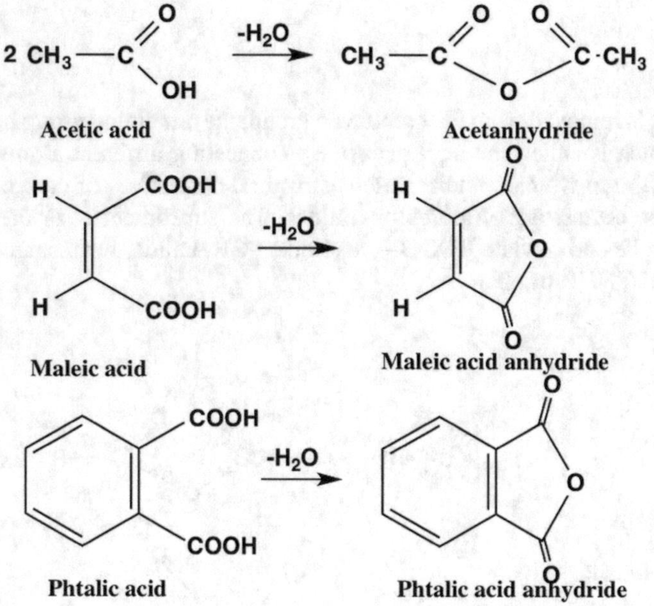

8.2   Esters, Nucleophilic Substitution on the Unsaturated Carbon Atom

The molecule of alcohol is a good nucleophile because of two electron lone pairs on the oxygen atom. In the reaction with carboxylic acid, the molecule of alcohol attacks via its oxygen atom the carbon atom of the carboxylic acid group. The mechanism is analogous to the previously discussed nucleophilic additions of aldehydes and ketones. The main difference between the two mechanisms is that the reaction involving carboxylic acid also involves the removing a molecule of water. The final product is **ester**.

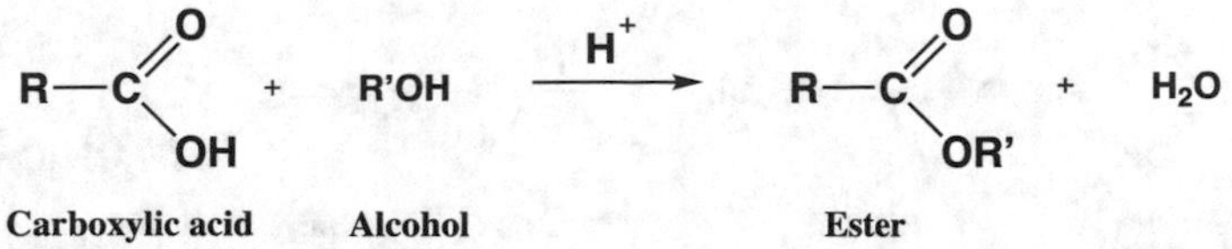

Carboxylic acid   Alcohol   Ester

ESTERIFICATION

The mechanism of this reaction which is called **esterification** is **nucleophilic substitution on the unsaturated carbon** (the carbon atom is unsaturated because it is bound with the double bond). In summary, esters are formed by the reaction of carboxylic acid and alcohol with the elimination of water molecule. However, removing the water molecule is possible only in acidic medium, so that in the preparation of esters a few drops of strong acid must be added (see the scheme below). The basic step in this mechanism is the protonation of the oxygen atom of the carboxylic group by a strong acid. Such protonated form is susceptible to the nucleophilic attack on the C=O group. The reaction intermediate is transformed into the ester molecule by removing the molecule of water:

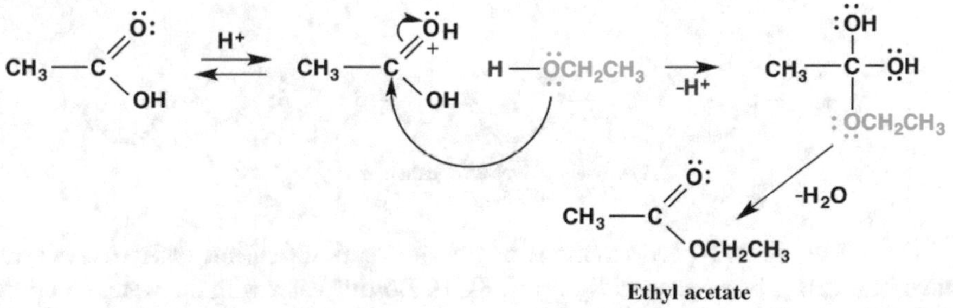

Ethyl acetate

Nomenclature of esters is relatively simple, because it is similar to the nomenclature of salts. For instance, the ester obtained from methanol and ethanoic acid (acetic acid) is called **methyl-ethanoate** (methyl-acetate). Esters are compounds that comprise a large number of structures depending on what alcohol and acid components are. Esters with small molecular mass are volatile substances, in most cases with pleasant odor. For instance, butyl-acetate is responsible for the odor of apples. In nature, some esters serve as pheromones for insects, for example **isoamyl-acetate**. Other kinds of esters can be large molecules as for instance waxes and fats where both, alcohol and acid components can be long-chained or complicated structures. These natural esters will be discussed in the chapter about lipids. In industry, esters are used for fabricating polymeric fibers for textile materials and plastic materials for various uses. The most common material is **polyethylene-terephthalate PET**, the ester prepared from terephthalic acid and ethylene glycol.

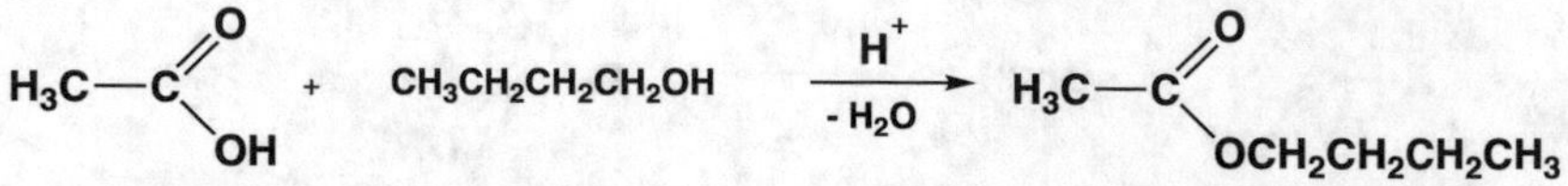

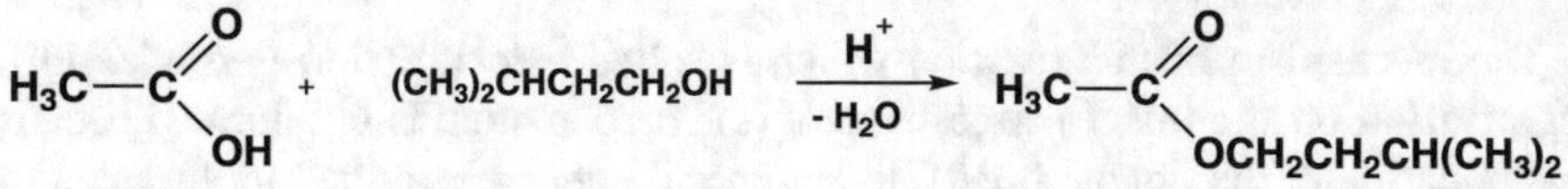

Ethanoic acid
Acetic acid   Butanol   Butyl ethanoat
                                                                Butyl acetet

Ethanoic acid   Isoamyl alcohol   Isoamyl acetate
Acetic acid

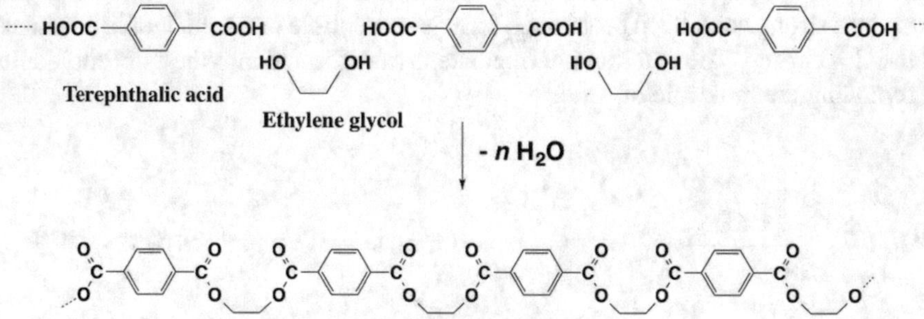

Terephthalic acid   Ethylene glycol

Polyethylene glycol-terephthalate
(PET)

One of the most typical reactions of esters is **saponification** or **hydrolysis of esters** by strong bases such as NaOH or KOH. Boiling ester with the water solution of KOH or NaOH yields the salt of carboxylic acid and the corresponding alcohol.

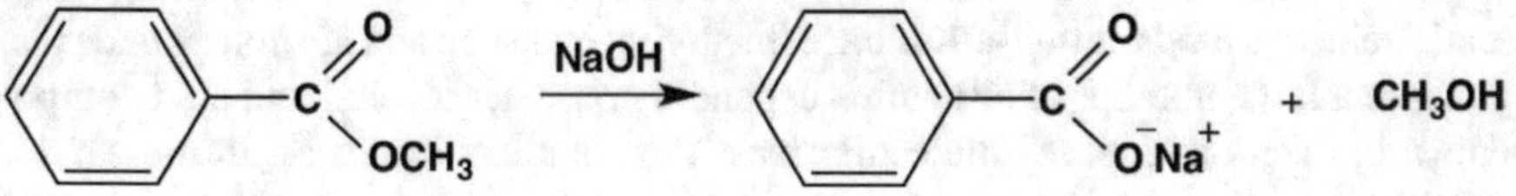

Methyl benzoate   Sodium benzoate   Methanol

Salts can be simply transformed into the carboxylic acid just by reacting with a strong acid. Remember the principle that strong acids can yield weak acids from their salts:

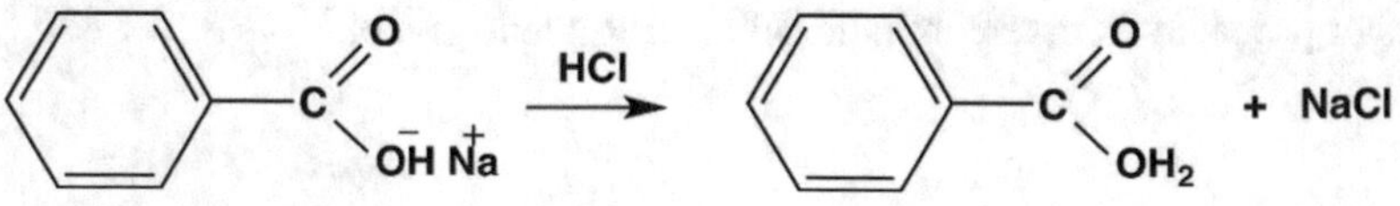

Sodium benzoate   Benzioc acid

Analogously with aldehydes and ketones, esters can be reduced by strong reducing agents such as LiAlH$_4$. Weaker reducing reagents such as NaBH$_4$ can reduce only aldehydes and ketones, but are unreactive towards esters. The reaction with LiAlH$_4$ must be performed in dry ether. The obtained intermediate cannot be isolated but is immediately hydrolyzed with diluted acid. The products are two alcohols, the first from the acid component and the second from the alcohol component of the ester precursor.

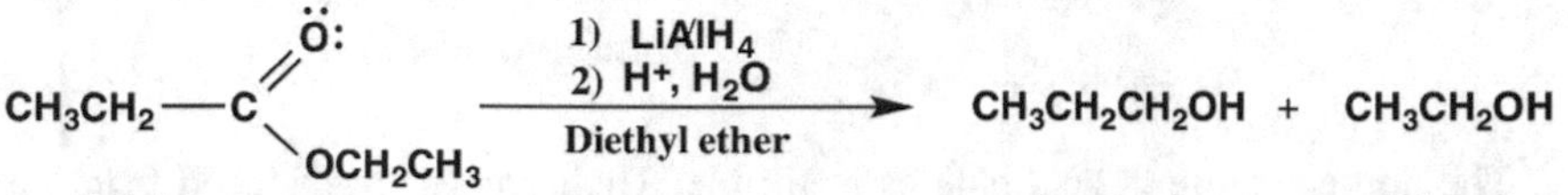

$$CH_3CH_2-C\overset{\overset{\ddot{O}:}{\|}}{\underset{OCH_2CH_3}{}} \quad \xrightarrow[\text{Diethyl ether}]{\substack{1)\ \text{LiAlH}_4 \\ 2)\ \text{H}^+,\ \text{H}_2\text{O}}} \quad CH_3CH_2CH_2OH \ + \ CH_3CH_2OH$$

Ethyl propanoate   **Propanol**   **Ethanol**

8.3 Acyl-Halides

Replacing the OH group in the molecule of carboxylic acid with halogen generates acyl-halides, the reactive compounds which are important intermediates in the organic synthesis of various other compounds, especially amides. The most frequently used acyl-halide is acyl chloride, the compound prepared from carboxylic acid and the special reagent called **thionyl chloride** SOCl$_2$. This reagent has unpleasant odor because it readily reacts with water yielding HCl and SO$_2$.

$$SOCl_2 \ + \ H_2O \ \longrightarrow \ SO_2 \ + \ 2HCl$$

Thionyl chloride reacts with carboxylic acids in the way as it is shown in the next scheme below. In the first step the reaction yields HCl and the intermediate benzoyl-chlorosulfate:

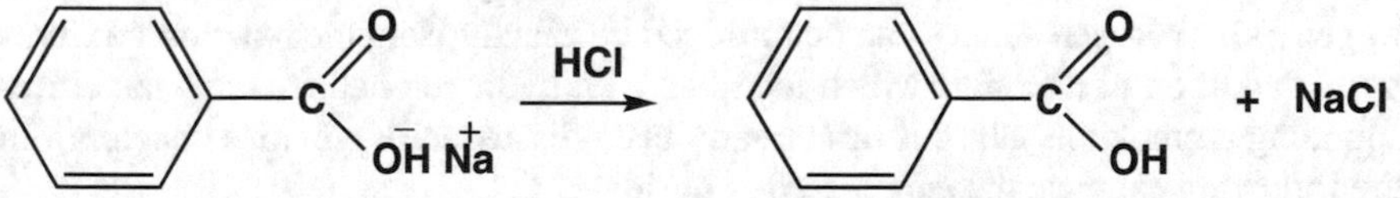

Sodium benzoate   **Benzioc acid**

Benzoyl-chlorosulfate immediately rearranges into the acyl chloride (benzoyl chloride in our example) and SO$_2$. This reaction is an interesting example of internal nucleophilic substitution. The internal nucleophile is the chlorine atom (it is electron-rich) which attacks the carbonyl carbon. The final products are benzoyl chloride and SO_2.

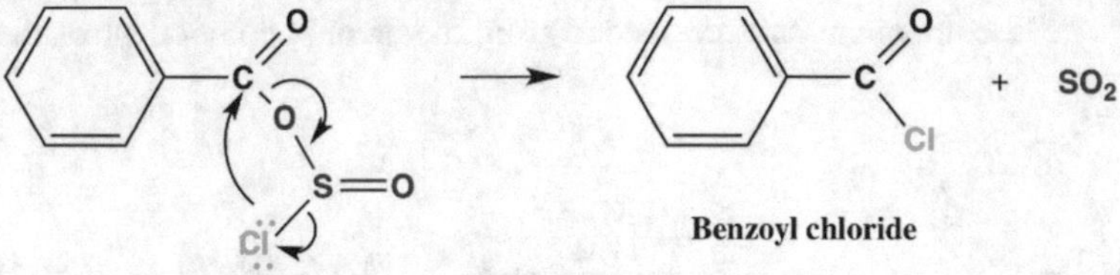

Benzoyl chloride

The nomenclature of acyl halides is simple. Their names are deduced from the name of the constituent acid with the suffix −**yl** or −**oyl** and the name of the halide.

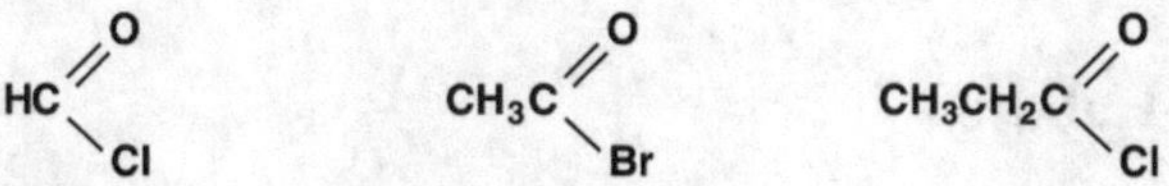

Methanoyl chloride   Ethanoyl bromide   Propanoyl chloride
(Formyl chloride)   (Acetyl bromide)

Like thionyl chloride, acyl-halides are sensitive to moisture and quickly hydrolyze into the corresponding carboxylic acid and HCl. This is an example of the nucleophilic substitution at the unsaturated carbon with H_2O as the nucleophile.

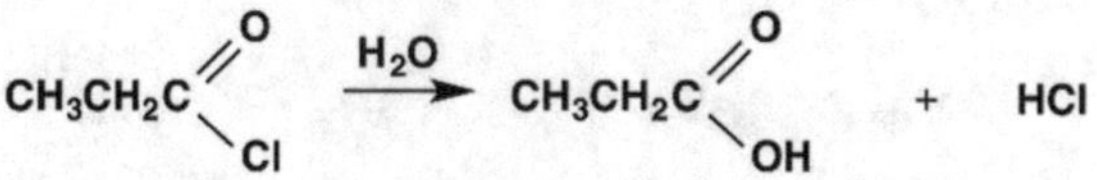

Propanoyl chloride   Propanoic acid

Besides OH− or water, other nucleophiles can also be involved in this type of the reaction. If the molecule of alcohol is the nucleophile, the products are ethers. The reaction is faster if HCl is removed immediately from the reaction mixture. In general, hydrogen halides can be removed efficiently from the reaction mixtures by the addition of an amine which together with hydrogen halide forms the corresponding ammonium salt (this has already been discussed in previous chapters). In the following example the amine used is pyridine.

(Pyridine)
HCl
Ethanoyl chloride
(Acetyl chloride)
Ethyl ethanoate
(Ethyl acetate)
Pyridine

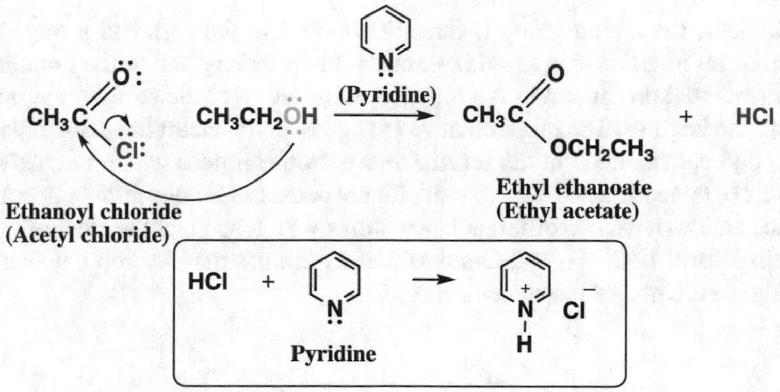

8.4 Amides

Perhaps the most important synthetic use of acyl halides is in the reactions with ammonia or with amines. The products are important natural compounds called **amides**. Depending on the structure of nucleophiles i.e., whether they are ammonia, primary or secondary amines, the products are **primary**, **secondary** or **tertiary amides**.

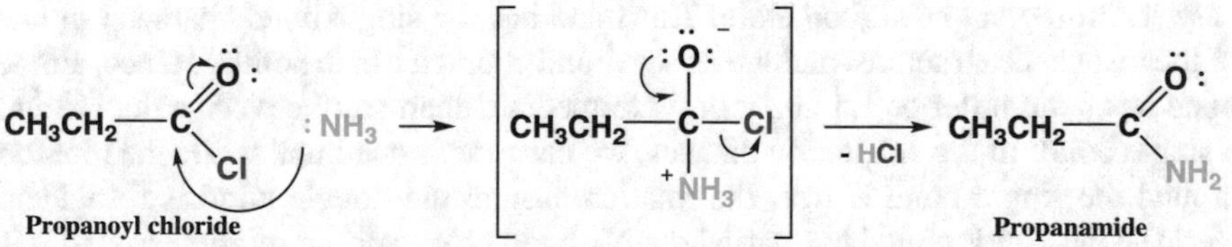

- HCl
Propanoyl chloride
Propanamide

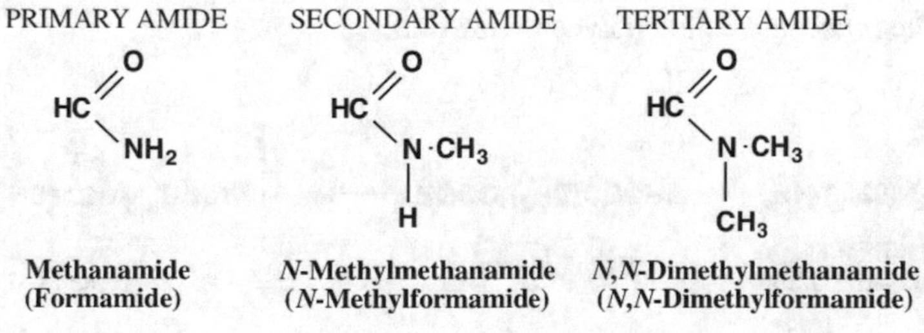

PRIMARY AMIDE
SECONDARY AMIDE
TERTIARY AMIDE
Methanamide
(Formamide)
N-Methylmethanamide
(N-Methylformamide)
N,N-Dimethylmethanamide
(N,N-Dimethylformamide)

In amides, the amino group is directly attached to the carbonyl group. Amide nomenclature includes simple suffix **–amide**. In secondary and tertiary amides, the substituents that are bound to the nitrogen atom are not labeled with the number, but with the letter N. The representative examples are N-methylmethanamide or N, N-dimethylmethanamide in the scheme above. Some amides which are derivatives of formic or of acetic acids also bear traditional names as **formamide** or **acetamide**.

Since amides have two different heteroatoms with lone pairs the electrons can be delocalized over both N– and O-atoms. Let us discuss the electron delocalization by using the method of resonance structures:

$$\left[\quad RC \overset{\displaystyle O:}{\underset{\displaystyle NH_2}{\diagdown}} \quad \longleftrightarrow \quad RC \overset{\displaystyle \overset{-}{O}:}{\underset{\displaystyle \overset{+}{NH_2}}{\diagdown}} \quad \right]$$

Two important conclusions can be drawn from the given resonance structures. First, the electron lone pair is not localized on the nitrogen atom as in saturated amines. We recall from the previous chapter that amines are less basic if the nitrogen lone pair is not localized on the N-atom. This was illustrated by an example where we compared the basicity of aniline and pyridine and we have demonstrated that aniline is a weaker base because its lone pair electrons are delocalized over the entire benzene ring. For the same reason, **amides** are neutral rather than basic.

The second very important conclusion is related to the nature of the covalent bond between nitrogen and carbon atom. This bond has the single bond character in one of the resonance structures and the double bond character in the other. Hence, amide bond has bond order of 1.5 i.e., it has intermediate character between a double and a single bond. In the chapter on alkanes, we have mentioned that while the rotation around the single bond is free, the rotation around the double bond is forbidden. Because the amides bond has partial double bond character the rotation around it is hindered. This evidence is important in the study of the stereochemistry of complex molecules with amide groups, such as polyamides and polypeptides.

One of the best-known polyamides is **nylon**, the substance which has been first produced in 1935. Nylon can be obtained from hexamethylenediamine and adipic acid by the procedure called **polycondensation**.

$$n\,H_2N(CH_2)_6NH_2 \;+\; n\,HOOC(CH_2)_4COOH \;\longrightarrow\; \left[NH(CH_2)_6NH\overset{\displaystyle O}{\overset{\displaystyle \|}{C}}(CH_2)_4C \right]_n$$

1,6-Diaminohexane   **Adipinic acid**   Nylon
(Hexamethylene diamine)   (Polyamide)

Many antibiotics and other pharmaceutical products have molecules with the amide groups. A large group of such compounds are antibiotics derived from penicillin (see the scheme below). These compounds are stable because the amide bond is difficult to hydrolyze. Penicillin will be discussed later in this book.

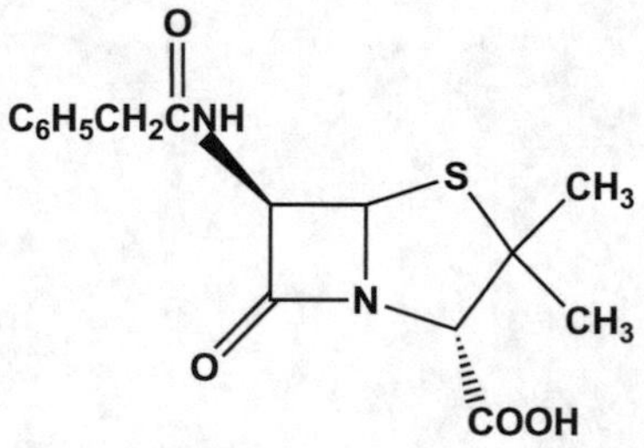

Penicillin G

(antibiotic)

Chapter 9
Electrophilic Substitutions

Reactions of Aromatic Compounds

In the chapter about hydrocarbons, we have discussed aromatic compounds, their stability and specific chemical behavior. The electrons in π orbitals are delocalized through the benzene ring and, although there are formal double bonds, aromatic compounds do not undergo standard addition reactions typical for alkenes. The distribution of π electrons in the ring is such that the highest electron density is above and below the plane of the aromatic ring. Such π-electron distribution is the reason that the benzene ring behaves like Lewis base which can donate electrons or react with strong Lewis acids. Protons and other positively charged particles are **electrophiles** because they tend to bind to species which are electron rich. Consequently, benzene and other aromatic molecules can be protonated by strong acids such as H_2SO_4.

The cation which is formed by protonation is stabilized because of the delocalization of electrons, as it is shown on the next scheme. For simplicity, these resonance structures can be replaced with a single formula where the region in which the electrons are delocalized is designated with dotted line and the + sign in the middle.

© The Author(s), under exclusive license to Springer Nature Switzerland AG 2022   127
H. Vančik, *Basic Organic Chemistry for the Life Sciences*,
https://doi.org/10.1007/978-3-030-92438-6_9

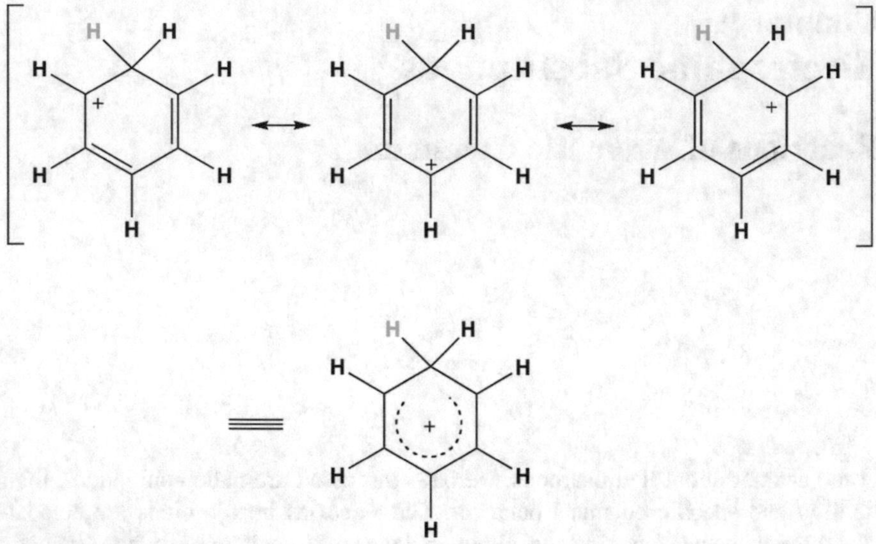

In general, the electrophile is labeled with E^+ and its addition to the benzene ring is analogous to protonation (see scheme below). If the cationic reaction intermediate loses proton, the electrophile remains attached to the ring. The final product has the structure in which hydrogen is replaced with the incoming electrophile and such reaction is called **electrophilic substitution** on the unsaturated carbon atom.

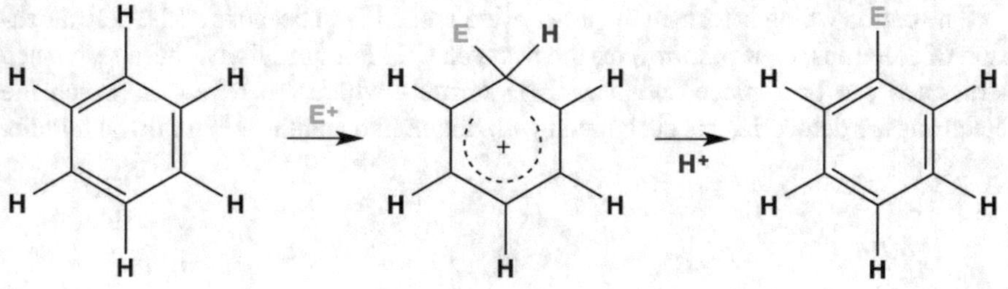

$$E^+ = \text{ELECTROPHILE}$$

If we wish to replace one hydrogen atom with some functional group or atom, this group or atom must be transformed into the cationic, electrophilic form. For instance, for the halogenation of benzene ring the halogen atom must appear in its electrophilic form called **halonium ion**. Since such ions are unstable in solution so they must be prepared in situ (immediately, without isolation) and in the form of the complex with a larger molecule. Such reactants with positive halogen appear in the reaction of halogens with metal halides such as $FeCl_3$, $FeBr_3$, or $AlCl_3$.

$$Br_2 + FeBr_3 \longrightarrow Br^+[FeBr_4]^- \quad \text{or} \quad Cl_2 + AlCl_3 \longrightarrow Cl^+[AlCl_4]^-$$

Bromonium or **chloronium** complexes which are shown in the examples above are good electrophiles and react with benzene by replacing the hydrogen atom with halogen.

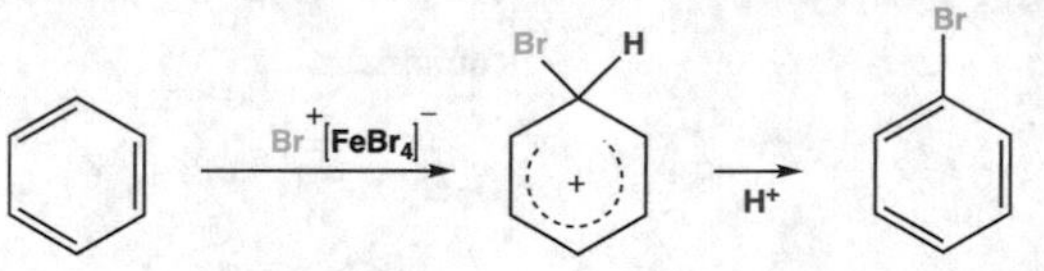

Bromobenzene

Alkyl or acyl groups can be introduced to benzene ring in an analogous way. These reactions are called **Friedel–Crafts alkylations** or **acylations**, respectively, in the honor of their discoverers **Charles Friedel** and **James M. Crafts**. The starting compound in alkylation is alkyl halide which together with the corresponding metal halide forms molecular complex that consists of carbocation and tetrahalometallic anion. For instance, ethyl bromide reacts with $FeBr_3$ forming the complex in which the ethyl group has cationic character and behaves as electrophile. This electrophile reacts with benzene molecule as it is shown in the following scheme.

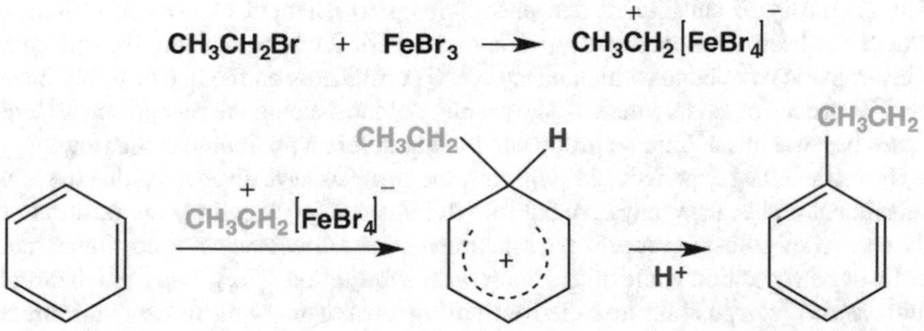

Ethylbenzene

In acylation the starting compound is the corresponding acyl-chloride:

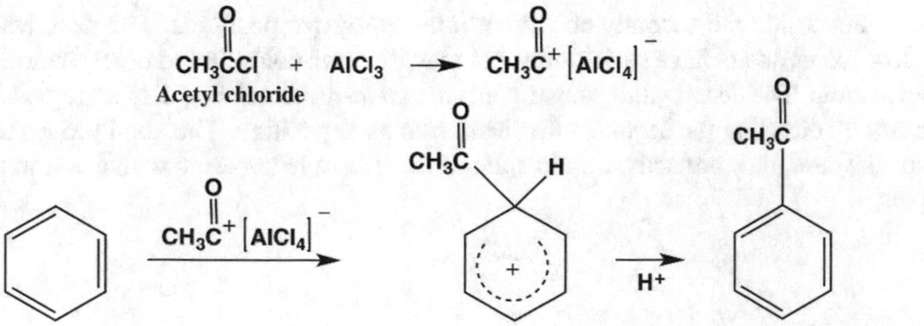

Phenylmethyl ketone

Reaction of nitric acid with sulfuric acid yields **nitronium ion** NO_2^+ which as an electrophile reacts with benzene and forms **nitrobenzene**.

$$HNO_3 \ + \ 2\,H_2SO_4 \ \longrightarrow \ NO_2^+ \ + \ H_3O^+ \ + \ 2\,HSO_4$$

Nitronium ion

Nitrobenzene

9.1 Substituent Effects in Electrophilic Aromatic Substitution

The chemistry of substituted benzenes belongs to the field of organic synthesis which has been developed during XIX century for the needs of the dye industry. The aromatic compounds which have several substituents on the benzene ring have been of special practical interest. The problem of introducing the second substituent arises because this additional group can bind at different positions on the ring.

From their long experience in synthesis, the chemists have discovered that the first substituent on benzene plays crucial role in influencing (**directing**) the position of the second substituent group. Some functional groups direct the next substituent into ortho or *para* position while others favor *meta* substitution. The groups which direct into *ortho* or *para* position are called **activating** substituents while those which direct into *meta* position are called **deactivating** groups. In principle, the term activating implies the donation of electrons to the benzene ring and deactivating has the opposite effect, i.e., the withdrawal of electrons from the ring.

The experience has shown that substituents with electron lone pairs on the atom directly bound to benzene are activating and they are also *ortho* or *para*-**directors** i.e., they guide the second substituent into *ortho/para* positions. The deactivating substituents do not have such atoms, but they can have double bond or an electronegative atom. The deactivating substituents are *meta*-**directors** i.e., display regioselectivity by directing the second substituent into *meta* position. The alkyl groups are a special case, they are activating in spite of the lack of heteroatom with electron lone pair.

ACTIVATING GROUPS
R_{ACT}

$\overset{..}{N}H_2$

Cl_2, $AlCl_3$

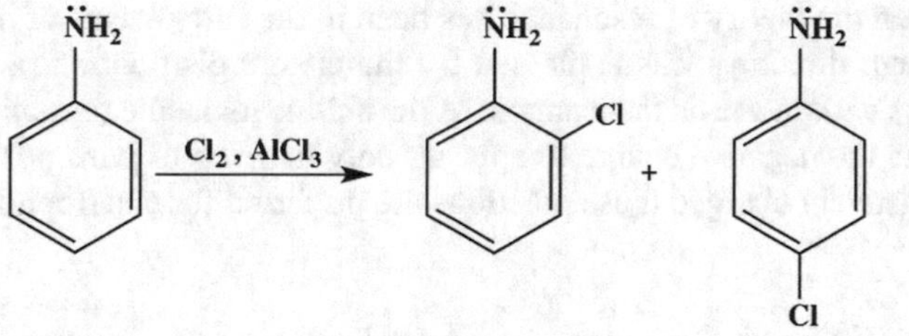

Aniline   o-Chloroaniline   p-Chloroaniline

DEACTIVATING GROUPS
R_{DE}

NO_2

Cl_2, $AlCl_3$

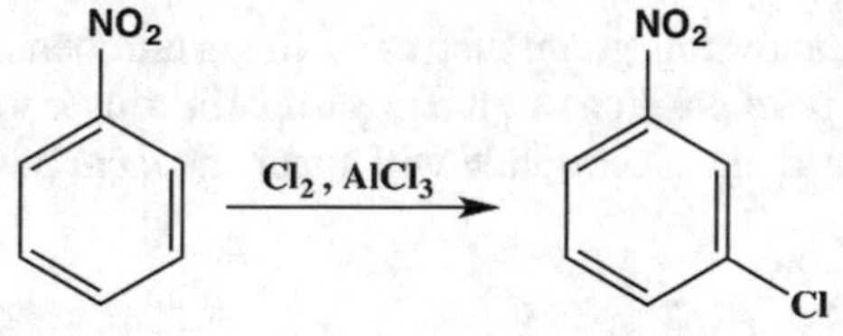

Nitrobenzene   m-Chloronitrobenzene

ACTIVATING GROUPS R_{ACT}	DEACTIVATING GROUPS R_{DE}
$-\overset{..}{N}H_2$	$-NO_2$
$-\overset{..}{N}HR$	$-CF_3$
$-\overset{..}{N}R_2$	$-SO_3H$
$-\overset{..}{\underset{..}{O}}H$	$-CN$
$-\overset{..}{\underset{..}{O}}R$	$-\overset{O}{\overset{\|}{C}}Cl$
$-\overset{..}{\underset{..}{X}}:$ for X = Cl, Br, I	$-\overset{O}{\overset{\|}{C}}OH$
$-R$ for R = alkyl	

These experimental observations have played crucial role not only in the development of the theory of reaction mechanisms, but also in the final solution of the problem of electronic structure of benzene. Historically, the explanation of the laws of the substituent direction appeared at two levels. During the first quarter of the last century, when the theory of resonance has been in the early stages of development, the substituent directing was explained by the charge distribution in the reactant molecule. As we can see on the example of the aniline molecule the activating –NH₂ group causes the negative charge to appears only in *ortho* or *para* positions. Since they are negatively charged these positions are preferred for electrophilic attack.

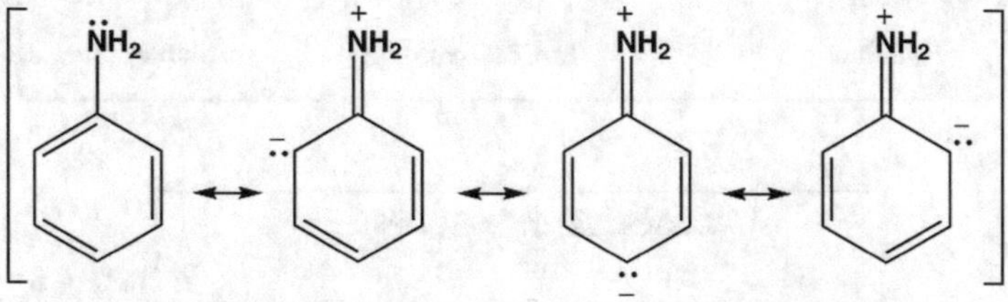

Conversely, the deactivating group such as – NO₂ in nitrobenzene induces positive charge in *ortho* and *para* positions so the electrophilic attack will not take place at these positions. Instead, the electrophile will attack the *meta* position.

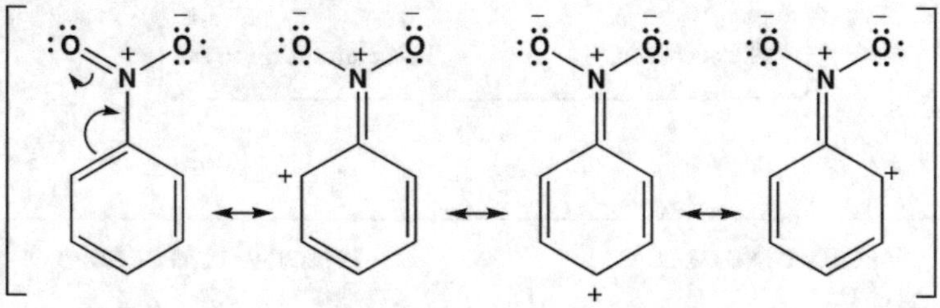

These initial ideas were inadequate to explain the orientation effect of all groups. The problem appeared with the methyl group in toluene which is activating, but for which no resonance structures could be constructed. The proper explanation of regioselectivity (the preference for certain substituent positions) only came with the development of the theory of reaction mechanisms by Robinson and Ingold. As we have seen at the beginning of this chapter, the electrophilic attack causes the formation of the reaction intermediate, carbocation. Since we know that the rate of chemical reaction depends mostly on the stability of the reaction intermediate, let us discuss the structure of the activated and deactivated cations.

In the following example, the activating substituent is methoxy group. If the electrophile attacks the *ortho* position the corresponding cation can be described by resonance structures shown in the following scheme.

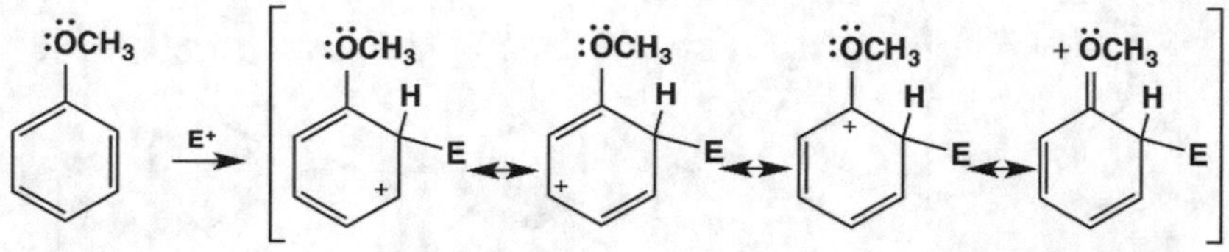

:ÖCH₃ → [:ÖCH₃ ↔ :ÖCH₃ ↔ :ÖCH₃ ↔ + ÖCH₃]

OXONIUM ION

The electron-donating property of the methoxy group is most evident in the last resonance structure describing the **oxonium ion** with positive charge on oxygen. The cation is stabilized by electrons transferred from the oxygen lone pairs. This is why all groups with electron lone pairs on the atom directly bound to the benzene ring are activating. Analogous resonance structures can also be written for aniline.

Methyl group, which does not have electron lone pairs is activating because it forms *inter alia* a stable tertiary carbocation (scheme below).

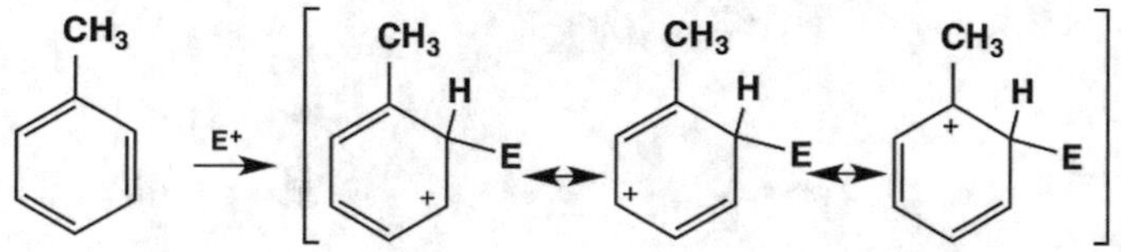

CH₃ → [CH₃ ↔ CH₃ ↔ CH₃]

TERTIARY
CATION

Electrophilic attack on *para* position leads to the formation of intermediate one of whose resonance structures is also stable tertiary carbocation. In conclusion, the activating substituents direct the next group into both *ortho* and *para* positions.

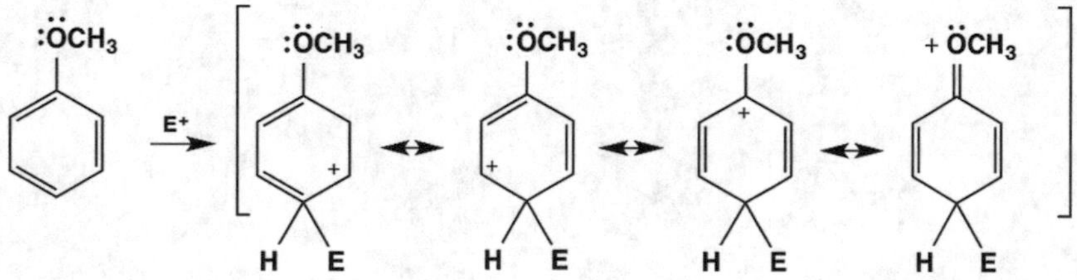

:ÖCH₃ → [:ÖCH₃ ↔ :ÖCH₃ ↔ :ÖCH₃ ↔ + ÖCH₃]

OXONIUM ION

The electrophilic attack on *meta* position would yield carbocation that cannot be stabilized by the activating group so this position is not favored. This can clearly be seen from the resonance structures. The carbon on which the activating group is bound always has double bond and the delocalization of electrons from the methoxy group into the ring is not possible. The corresponding carbocation is not stabilized and activating groups do not support the formation of *meta*-products.

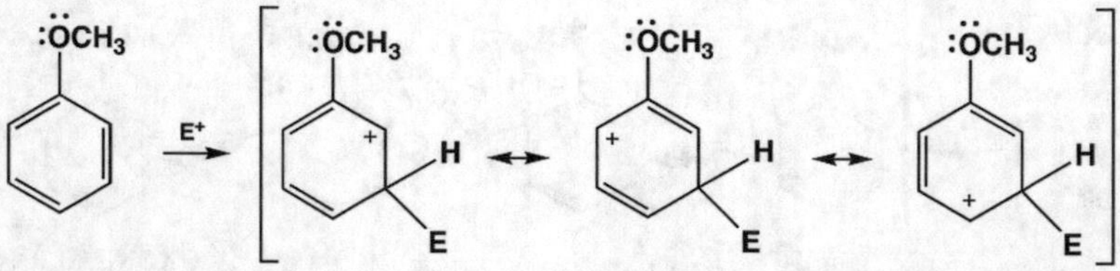

The deactivating groups withdraw electrons from the ring and the binding of electrophile will yield intermediate in which the resonance structures can be written as shown in the scheme below. In *ortho* or *para* position, the electrophile induces the formation of intermediates in which two positive charges are located next to each other (labeled with red circle in the scheme), which is energetically unfavorable. However, such situation is avoided if electrophile is bound in *meta* position. Hence the deactivating group is always *meta*-director.

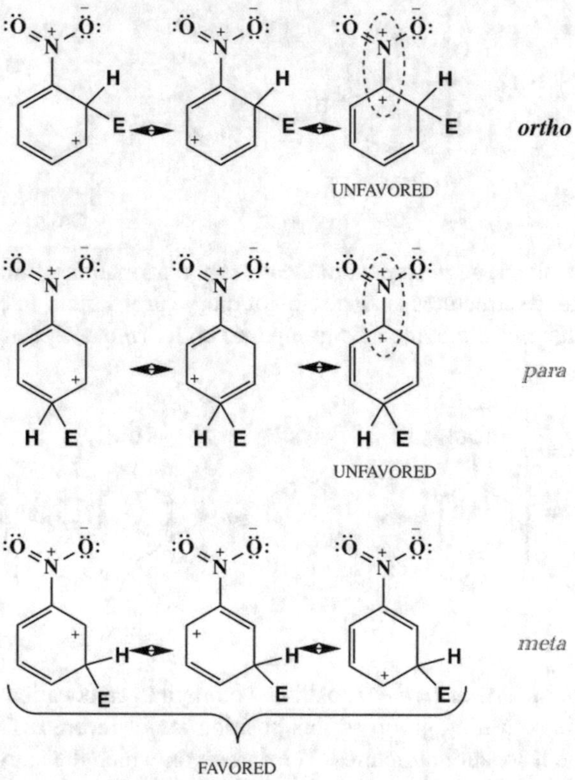

The deactivating effect of the CF_3 group is the consequence of strong electron withdrawing of electron density from the benzene ring via inductive effect of the three fluorine atoms.

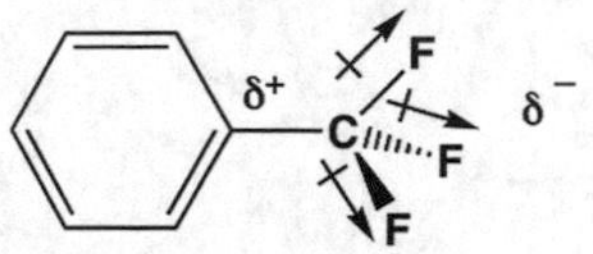

The typical example of the substituent-director effect (regioselectivity) is the preparation of the well-known explosive **trinitrotoluene** (**TNT**) by nitration of toluene. Methyl group is activating and directs nitro substituents into *ortho* and *para* positions.

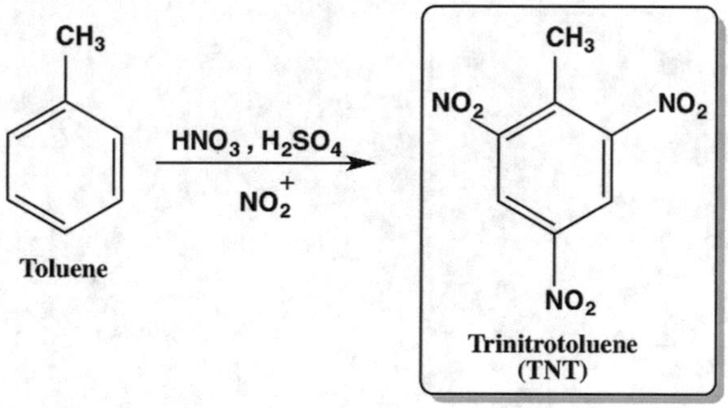

Chapter 10
Pericyclic Reactions

In the first chapters of this book, we have mentioned polycyclic compounds. Some of the reactions in which polycyclic compounds are formed have been known for more than a hundred years, but their mechanisms remained puzzling for the long period of time. Colloquially, these reactions were even called "reactions without the mechanism". Today, plenty of such reactions are known under the name **pericyclic reactions**. Here we will represent the main principles of the three most common pericyclic reaction types, **cycloadditions, electrocyclizations**, and **sigmatropic rearrangements**. The oldest known cycloaddition reaction is **Diels–Alder addition,** named in the honor of **Otto Diels** and **Kurt Alder**. In this reaction two alkene molecules are condensed so that two new σ-bonds are formed. The illustrative examples are the reactions in which cyclohexene and norbornene skeletons are formed:

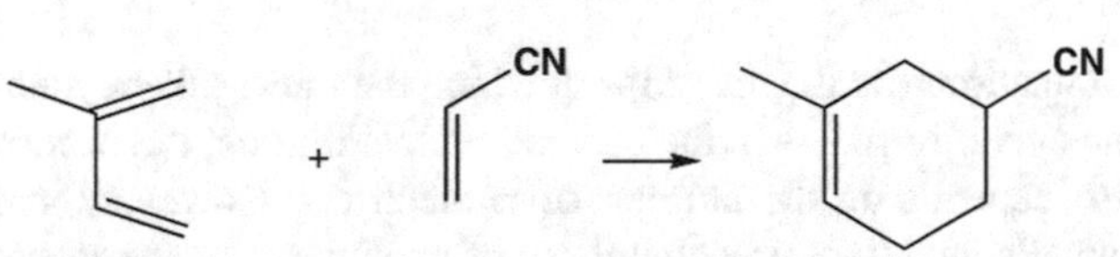

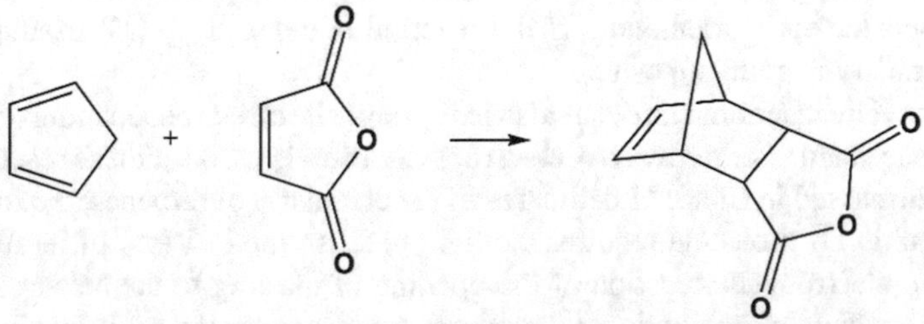

The occurrence of these reactions strongly depends on the electronic structures of reactants, especially on the number of π electrons. The reactions are favored if one of the reactants has 4 π electrons and the other has 2 π electrons (4n + 2 π electrons

H. Vančik, *Basic Organic Chemistry for the Life Sciences*,
https://doi.org/10.1007/978-3-030-92438-6_10

addition). These 6 π electrons are rearranged in the product into two newly formed σ bonds (labeled in red in the scheme) and one new π bond.

4 π Electrons   2 π Electrons

Other combinations of reactants such as 2 π electrons with 2 π electrons or 4 π electrons with 4 π electrons do not react under thermal conditions, but readily react if they are irradiated with the visible or ultraviolet light. These rules of reactivity in cycloadditions were discovered by **Robert B. Woodward** and **Roald Hoffmann** on the basis of quantum–mechanical analysis and the symmetry properties of molecular orbitals. These rules are known as **Woodward-Hoffmann rules**.

Electrocyclizations are the reactions in which the linear alkene molecule is transformed in the cyclic form. As in the case of cycloadditions, the occurrence of these reactions also depends on the number of π electrons. However, here the number of π electrons dictates the stereochemistry of products. For the reactants with 4 π electrons the thermally induced cyclization yields the product with the *trans*-oriented substituent (CH$_3$ groups in the Fig. 10.1), but for the reactants with 6 π electrons the products have *cis* orientation. If the reaction is induced by UV irradiation the stereochemistry is quite opposite.

Electrocyclic reactions can occur also in the opposite direction: opening of the ring to the alkene chain. The term **retro-electrocyclization** is used for these reactions. As it will be discussed in Chap. 11 dedicated to the chemistry of terpenes, the formation of the vitamin D$_3$ molecule requires sunlight because the last step of its formation is the retro-electrocyclic reaction of the opening of the ring to the hexatriene form (Fig. 10.2), which, after additional rearrangement produces the molecule in its final form. This last reaction is also a sort of pericyclic reactions called **sigmatropic rearrangements**.

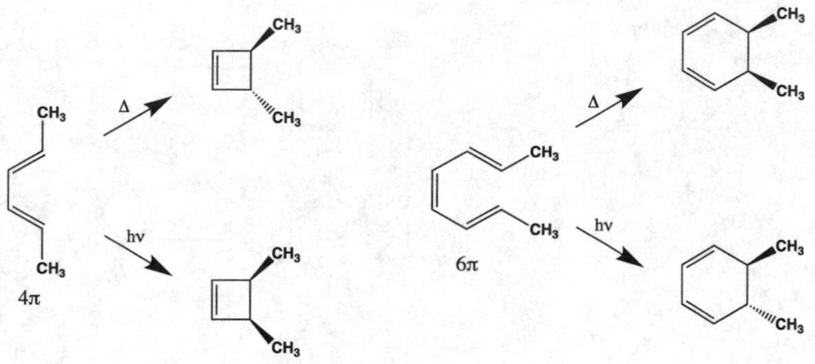

Fig. 10.1 Stereochemistry of electrocyclizations

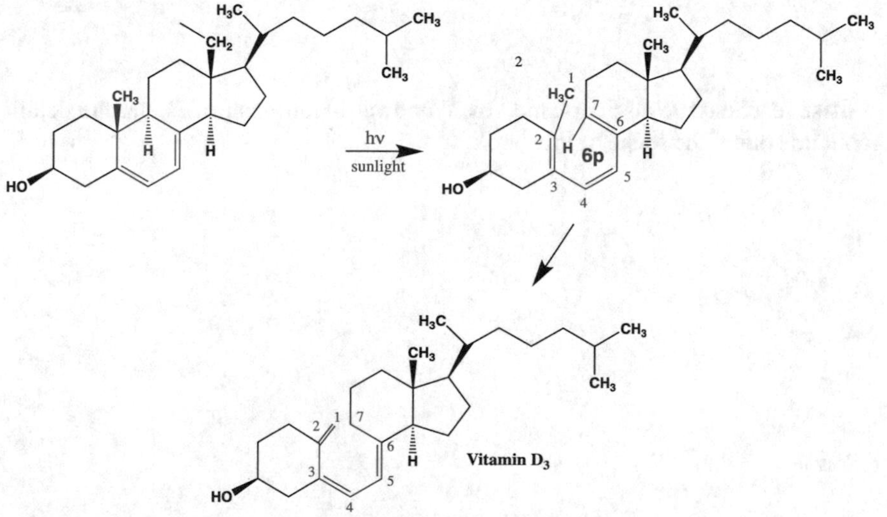

Fig. 10.2 Pericyclic reactions in the formation of vitamin D_3. The first reaction represents the stereochemistry of the retro-electrocyclization: the opening of the hexagonal ring to the 6 π electron system (labeled in red). The second reaction is the [1,7] hydrogen shift (the sigmatropic rearrangement)

Sigmatropic rearrangements include the shift of the hydrogen atom through the carbon atom chain, and the double bonds are shifted in neighboring positions. Depending on the number of C-atoms included in the sigmatropic rearrangements, the hydrogen can be shifted from C_1 to C_3 ([1,3] shift), from C_1 to C_5 ([1,5] shift), C_1 to C_7 ([1,7] shift), etc. (Fig. 10.3). The last step in the formation of vitamin D_3 represented in the Fig. 10.2 is the [1,7] sigmatropic hydrogen shift. As it can be recognized from the red colored fragment on the formula, the H-atom is from the CH_3 group transferred for 7 C-atoms. Thus, this CH_3 group is converted to $= CH_2$, and the three double bonds are shifted to neighboring positions. Stereochemistry

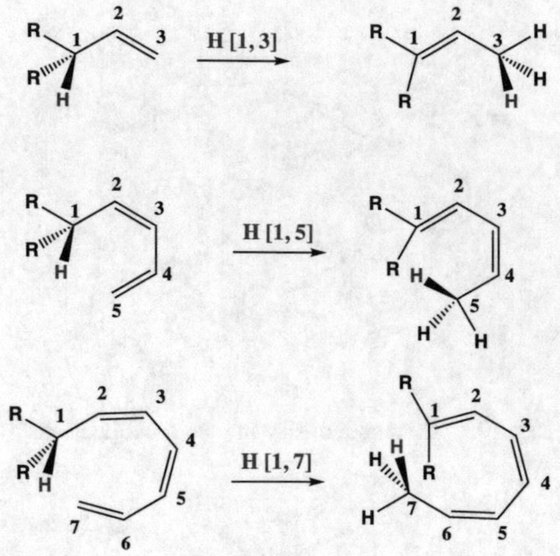

Fig. 10.3 Basic sigmatropic rearrangements

of these reactions are also explained by Woodward-Hoffmann rules, but the details about it is out of the scope of this book.

Chapter 11
Organic Natural Products

The knowledge of basic structural and dynamic concepts of organic chemistry allows us to examine some aspects of the chemistry of living organisms. The branch of organic chemistry that deals with compounds isolated from the living organisms is called **chemistry of natural products**. Basic knowledge of this field of chemistry is of fundamental importance for the understanding biochemistry.

In principle, the natural products can be divided in two categories, **primary** and **secondary metabolites**. Primary metabolites are compounds that are necessary for the functioning of the organism and with minor variations they are common to all living organisms. This group comprises amino acids, carbohydrates, lipids and nucleotides. Some organic compounds that are characteristic of a specific biological genus belong to secondary metabolites. Good examples of secondary metabolites are alkaloids, some terpenes, flavonoids, etc. These classes of compounds are not common to all organisms. The literature definition states that the **chemistry of natural products** has main interest in secondary metabolites while the science interested in primary metabolites is called **bioorganic chemistry**. Bioorganic chemistry is closely related to **biochemistry** which investigates the chemical systems in living organisms.

Our discussions about the natural products will begin with the primary metabolites. Since the organic compounds in nature always exist in complex mixtures, chemists developed special laboratory methods for their purification and isolation. One of earliest such methods is extraction, the technique that is based on different solubility of substances. Because similar substances dissolve in similar solvents, most organic substances will be soluble in organic solvents: polar substances in polar solvents, nonpolar in nonpolar solvents. In contrast, ionic compounds are well soluble in water. Amongst primary metabolites, the most water-soluble compounds are amino acids and polar carbohydrates. A large group of structurally diverse natural products soluble in nonpolar solvents are lipids which will be discussed at the end of this chapter.

11.1   Amino Acids and Peptides

Physiologically most important components of living beings are proteins, the carriers of basic physiological and biochemical functions. Their molecules are extraordinarily complex and varied but in spite of this complexity their structures are built from the only twenty relatively simple molecules—the **amino acids**. Before investigating these biologically important amino acids which are products of hydrolysis of proteins, we shall describe the structure and properties of this class of compounds in general.

The molecules of amino acids have two functional groups with the opposite properties, the basic amino group and the acidic carboxyl group. Because amino group can be bound to different carbon atoms on the hydrocarbon chain, the names of these compounds are derived from the position of this functional group. However, for this class of compounds the traditional nomenclature in which C-atoms are labeled not by numbers, but by letters of Greek alphabet is still in use. In addition, the letter α does not correspond to the carbon atom labeled with number 1, but to the atom with the number 2. In this nomenclature there are α, β, γ **amino acids**, etc.

γ β α	γ β α	γ β α
4 3 2 1	4 3 2 1	4 3 2 1
$CH_3CH_2CHCOOH$	CH_3CHCH_2COOH	$H_2NCH_2CH_2CH_2COOH$
|	|	
NH_2	NH_2	
2-Aminobutanoic acid	3-Aminobutanoic acid	4-Aminobutanoic acid
α-Aminobutanoic acid	β-Aminobutanoic acid	γ-Aminobutanoic acid

All twenty amino acids obtained from natural proteins belong to **α-amino acids** and can be represented by the general formula $H_2NCHRCOOH$. They differ only in the substituent R.

Amino acids are amphoteric compounds; carboxylic groups are giving them acidic and amino groups basic character. Their amphoteric behavior is shown in the next scheme.

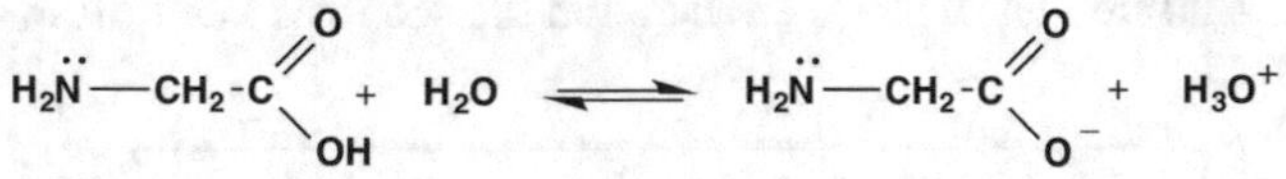

AMINO ACID AS ACID

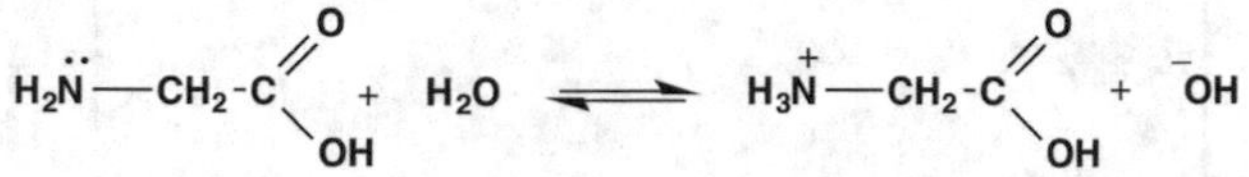

AMINO ACID AS BASE

Amino acids are well soluble in water, because they can appear in the form of dipolar ion called by the German word Zwitterion. Formally, this ion can be considered as the product of self-protonation: the amino group is protonated by the proton that comes from the carboxyl group.

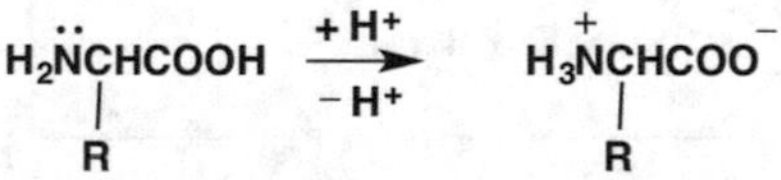

DIPOLAR ION

ZWITTERION

When R group is neutral the pH of the solution of such amino acid would be approximately 7, i.e., this amino acid is neutral. However, R groups can be neutral, acidic or basic. For instance, in aspartic acid (see the following scheme) the group R has carboxylic functional group. Consequently, pH of the water solution of aspartic acid must be lower than 7 and it is actually 2.95. On the other hand, in the amino acid lysine the R group contains amino group and the compound behaves as a base. In water solution, lysine has pH equal to 9.8. The pH value of the amino acid in water solution depends on the functional group R and is called **isoelectric point** which is labeled with the symbol **pI**.

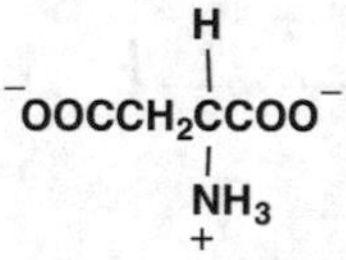

Aspartic acid

pI = 2,95

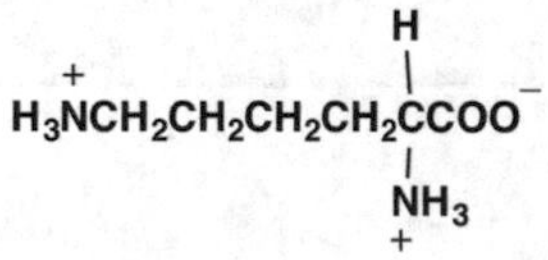

Lysine

pI = 9,8

In the following schemes all 20 amino acids are given and classified into neutral, basic and acidic forms and their pI values are represented. Since in biochemistry we handle complex protein molecules consisting of thousands of amino acids, the writing of exact formulas becomes unpractical, so every amino acid is labeled by three letters. All three-letter labels are listed in the schemes that follow.

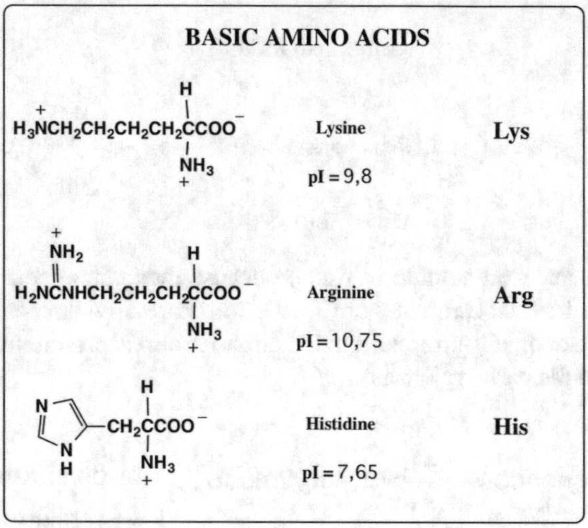

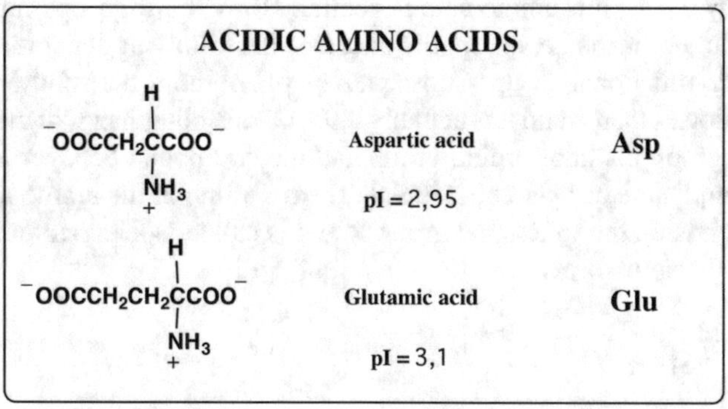

BASIC AMINO ACIDS

$$\overset{+}{H_3}NCH_2CH_2CH_2CH_2\overset{\overset{H}{|}}{\underset{\underset{+}{NH_3}}{C}}COO^-$$

Lysine   Lys

pI = 9,8

$$H_2N\overset{\overset{+}{NH_2}}{\overset{\|}{C}}NHCH_2CH_2CH_2\overset{\overset{H}{|}}{\underset{\underset{+}{NH_3}}{C}}COO^-$$

Arginine   Arg

pI = 10,75

Histidine   His

pI = 7,65

ACIDIC AMINO ACIDS

$$^-OOCCH_2\overset{\overset{H}{|}}{\underset{\underset{+}{NH_3}}{C}}COO^-$$

Aspartic acid   Asp

pI = 2,95

$$^-OOCCH_2CH_2\overset{\overset{H}{|}}{\underset{\underset{+}{NH_3}}{C}}COO^-$$

Glutamic acid   Glu

pI = 3,1

NEUTRAL AMINO ACIDS WITH POLAR GROUPS R

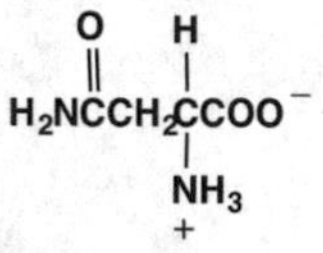

$$\underset{\overset{\displaystyle \|}{O}}{H_2NCCH_2}\overset{\overset{\displaystyle H}{|}}{\underset{\underset{+}{NH_3}}{C}}COO^-$$

Asparagine   **Asn**

pI = 5,40

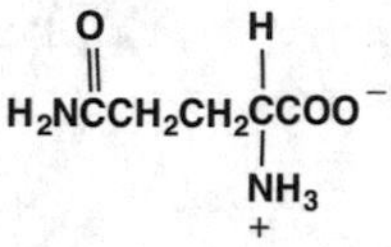

$$\underset{\overset{\displaystyle \|}{O}}{H_2NC}CH_2CH_2\overset{\overset{\displaystyle H}{|}}{\underset{\underset{+}{NH_3}}{C}}COO^-$$

Glutamine   **Gln**

pI = 5,65

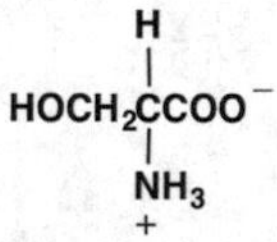

$$HOCH_2\overset{\overset{\displaystyle H}{|}}{\underset{\underset{+}{NH_3}}{C}}COO^-$$

Serine   **Ser**

pI = 5,70

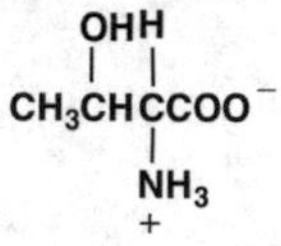

$$CH_3\overset{\overset{\displaystyle OH}{|}}{C}H\overset{\overset{\displaystyle H}{|}}{\underset{\underset{+}{NH_3}}{C}}COO^-$$

Threonine   **Thr**

pI = 5,60

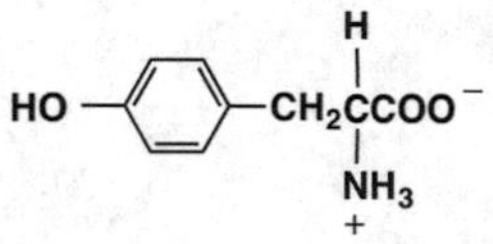

$$HO-\!\!\bigcirc\!\!-CH_2\overset{\overset{\displaystyle H}{|}}{\underset{\underset{+}{NH_3}}{C}}COO^-$$

Tyrozine   **Tyr**

pI = 5,70

$$HSCH_2\overset{\overset{\displaystyle H}{|}}{\underset{\underset{+}{NH_3}}{C}}COO^-$$

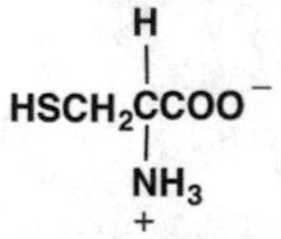

Cysteine   **Cys**

pI = 5,15

NONPOLAR AMINO ACIDS WITH NONPOLAR SIDE GROUPS R

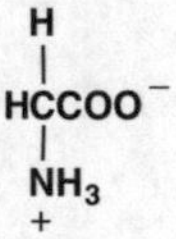

H $\mid$ $HCCOO^-$ $\mid$ NH_3 $+$	Glycine pI = 10,75	**Gly**
H $\mid$ CH_3CCOO^- $\mid$ NH_3 $+$	Alanine pI = 6,15	**Ala**
H $\mid$ $(CH_3)_2CHCCOO^-$ $\mid$ NH_3 $+$	Valine pI = 6,00	**Val**
H $\mid$ $(CH_3)_2CHCH_2CCOO^-$ $\mid$ NH_3 $+$	Leucine pI = 6,00	**Leu**
CH_3 H $\mid$ $\mid$ $CH_3CH_2CH-CCOO^-$ $\mid$ NH_3 $+$	Isoleucine pI = 6,05	**Ile**
H $\mid$ $CH_3SCH_2CH_2CCOO^-$ $\mid$ NH_3 $+$	Methionin pI = 5,70	**Met**
$\overset{+}{N}H_2$ COO^-	Proline pI = 6,30	**Pro**
H $\mid$ $-CH_2CCOO^-$ $\mid$ NH_3 $+$	Phenylalanine pI = 5,75	**Phe**
H $\mid$ $-CH_2CCOO^-$ $\mid$ NH_3 $+$	Tryptophan pI = 5,95	**Trp**

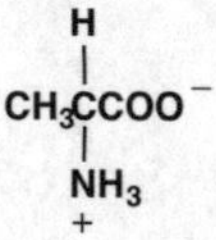

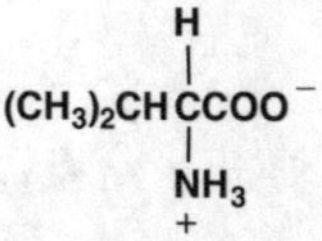

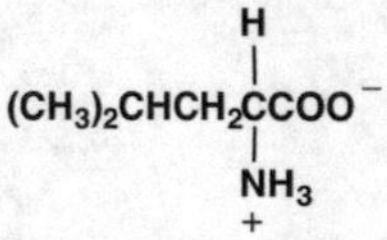

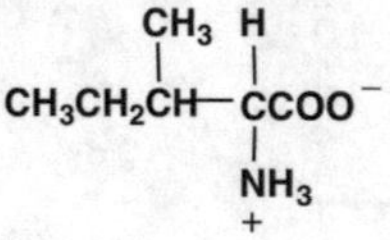

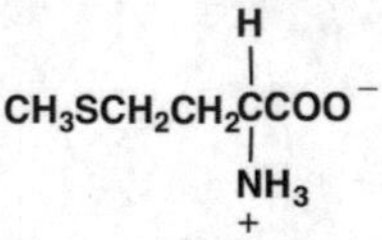

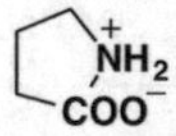

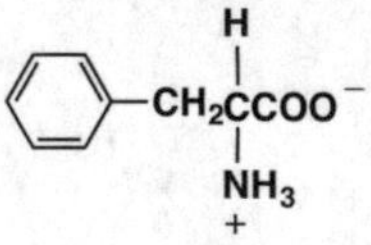

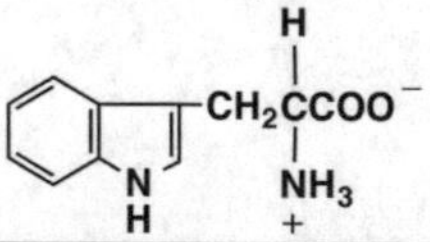

Amino acids valine, leucine, isoleucine, methionine, phenylalanine, tryptophane, threonine, lysine, arginine and histidine are called **essential**, because they are necessary components of the food of mammals. Namely, mammalian organisms are not able to synthesize amino acids in this group.

In all the listed amino acids, with the exception of glycine, the α-carbon is bound to four different substituents hence it is the stereogenic center. From this it follows that every amino acid can appear in the form of two enantiomers. In the following example, both the enantiomers of alanine are represented together with their absolute configurations. However, enantiomers of amino acids can also be represented by traditional notation of chiral molecules that is called the **relative configuration**. This nomenclature for configuration was proposed **Emil Fischer** in XIX century for the representation of stereochemistry of carbohydrates.

For the determination of relative configuration, the molecule is oriented so that the chiral carbon atom is in the center of the cross, the substituents positioned at the top and bottom arms of the cross lie bellow the plane of drawing and the groups at the left and right arm are above the plane of the drawing. Substituents H and NH_2 are always either the left or the right arm and the carboxylic group is on the top arm of the cross. These representations are called **Fischer formulas**. If the NH_2 group is on the left arm of the cross, the configuration is designated L-**configuration**. Conversely, if NH_2 is on the right arm the molecule has D-**configuration**. Relative and absolute configurations of alanine are presented in the next scheme. It must be pointed out that absolute and relative configurations do not correlate: (R) is not always D and (S) is not always L.

It is important to mention that **all twenty amino acids obtained from living organisms have the same L relative configuration**. Although science does not have the full explanation for this observation such stereoselectivity supports the notion that chemical and later biological evolution started from the same enantiomer.

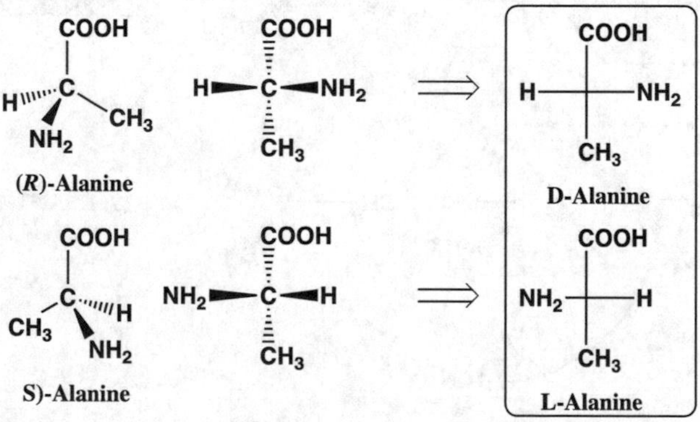

FISCHER`S FORMULAS

The laboratory synthesis of amino acids is fundamentally different from the way these compounds are synthesized in the living organisms (**biosynthesis**). The traditional synthesis of amino acids is known as **Strecker synthesis** (in the honor of **Adolph Strecker**). This method is based on the nucleophilic attack on the carbonyl group; the reaction which we have already discussed in this book. In this reaction of aldehyde with ammonium chloride and sodium cyanide the product is aminonitrile. By boiling with strong acid, this aminonitrile is transformed into the corresponding amino acid.

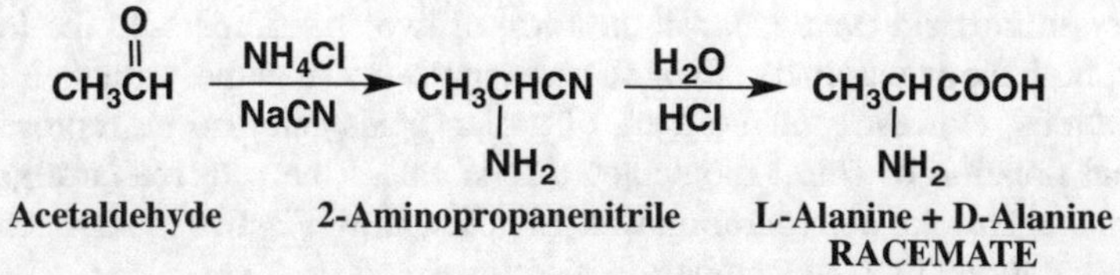

$$CH_3\overset{\overset{O}{\|}}{C}H \xrightarrow[\text{NaCN}]{\text{NH}_4\text{Cl}} CH_3\underset{NH_2}{C}HCN \xrightarrow[\text{HCl}]{\text{H}_2\text{O}} CH_3\underset{NH_2}{C}HCOOH$$

Acetaldehyde   **2-Aminopropanenitrile**   **L-Alanine + D-Alanine**
                                                              RACEMATE

From the chemical equations above it is clear that the laboratory synthesis is not a "clean" reaction, because it yields a mixture of both enantiomers: L-alanine and D-alanine i.e. the racemate. The appearance of racemate can be explained from the knowledge of the reaction mechanism of nucleophilic addition to the carbonyl group. In the carbonyl group the carbon atom is bound by double bond and lies with all three of its substituents in the same molecular plane. The CN^- group as a nucleophile can attack the carbonyl carbon with the same probability from either side of the molecular plane. Consequently above-the-plane attack yields (R) enantiomer and below-the-plane attack gives the (S) enantiomer.

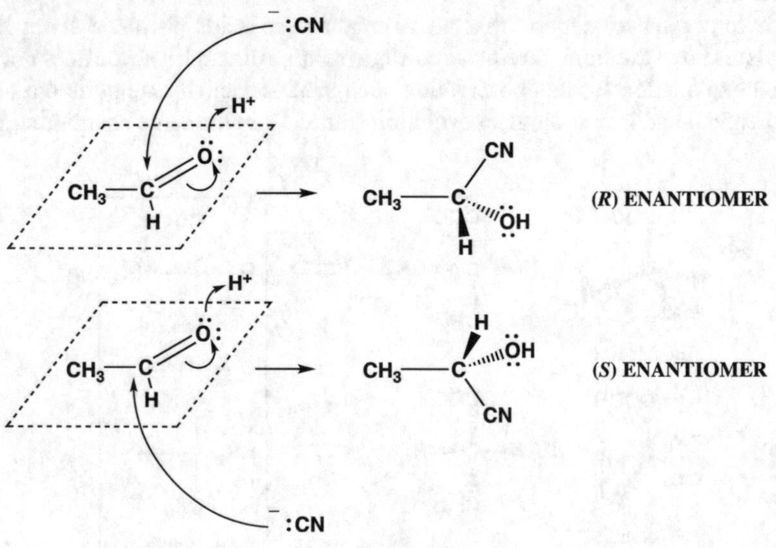

Biosynthesis of amino acids in living organisms is a more complex process that involves biocatalysts, enzymes. Such biocatalytic process is stereoselective what means that it leads to only one enantiomer as a product, the enantiomer always having L-configuration.

One of known mechanisms of biosynthesis is shown in the next scheme. The enzyme **aminotransferase** transfers the amino group from the precursor amino acid to the **α-ketoacid**, which is rearranged into the product amino acid. In this process, the precursor amino acid is transformed into the corresponding **α-ketoacid**. It must be pointed out that besides transferring the amino group, the enzyme aminotransferase also preserves the stereochemistry: L-precursor amino acid leads to the formation of the L-product amino acid. This mechanism presented below is the representation of the biosynthesis of the L-glutamic acid.

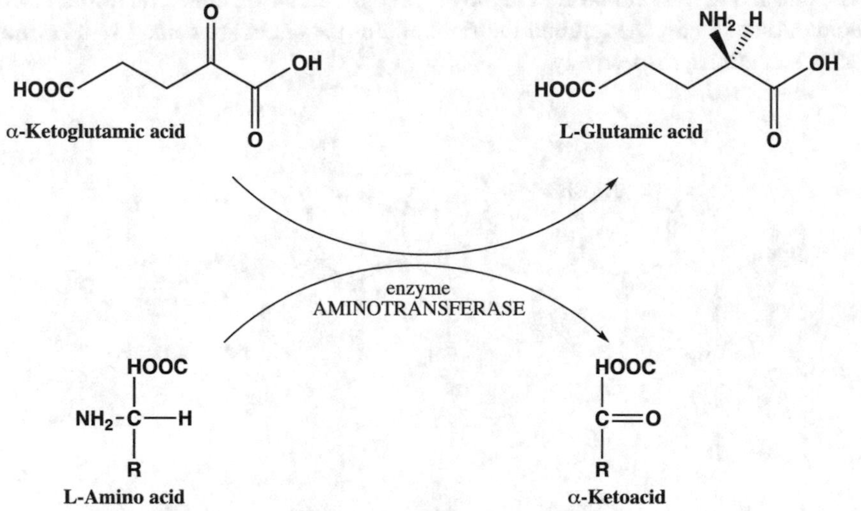

The most important property of amino acids is their ability to form dimers, oligomers and polymers that are called **peptides** and **polypeptides**. The mechanism of the formation of peptides and polypeptides is complex and it will not be discussed in this book. Let us only mention, that the **peptide bond** is formed by the connecting the amino group of one amino acid with the carboxyl group of the other amino acid followed by elimination of water.

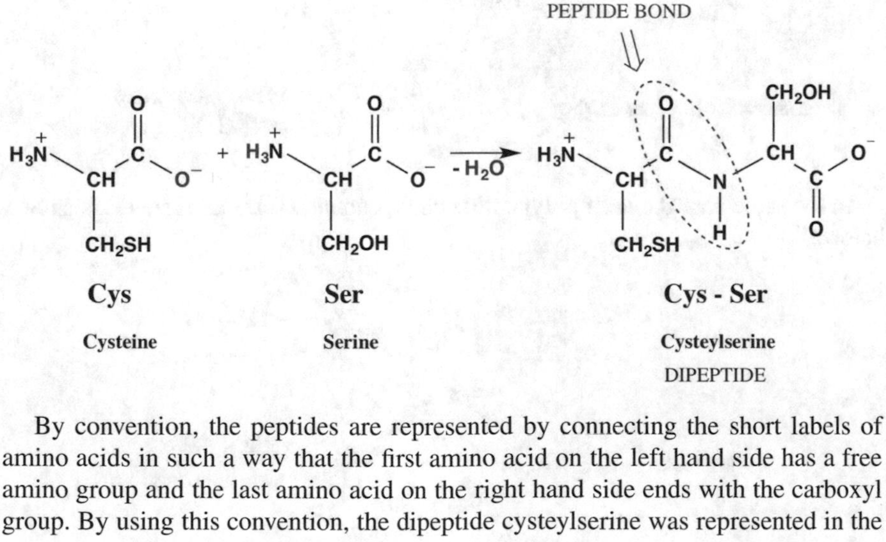

By convention, the peptides are represented by connecting the short labels of amino acids in such a way that the first amino acid on the left hand side has a free amino group and the last amino acid on the right hand side ends with the carboxyl group. By using this convention, the dipeptide cysteylserine was represented in the scheme above. In reality, the **peptide bond** is the **amide group** because there is an amino group directly bonded to the carbonyl group. Large number of amino acids can be connected into polypeptides that are represented in the same way as dipeptides. The first and the last amino acid in the polypeptide chain are called **terminal amino acids**. Accordingly, the left and the right end of the chain are named **N-terminus**, and **C-terminus,** respectively.

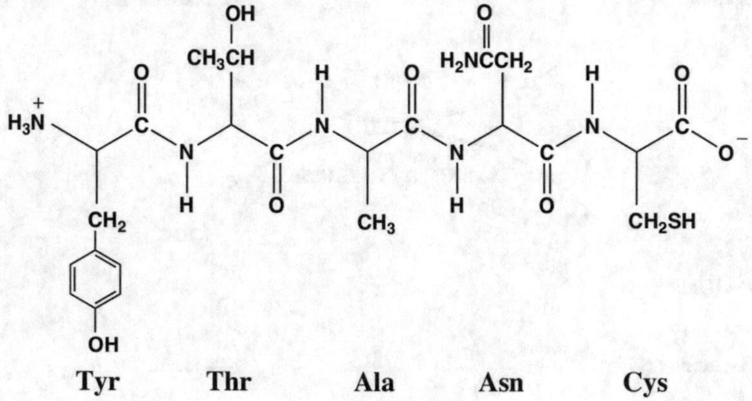

<table>
<tr><td>Tyr</td><td>Thr</td><td>Ala</td><td>Asn</td><td>Cys</td></tr>
</table>

N-terminus   **Tyr-Thr-Ala-Asn-Cys**   **C-terminus**

Longer polypeptide chains can be interconnected by the formation of the **disulfide bond** as shown for the two polypeptides in the scheme below.

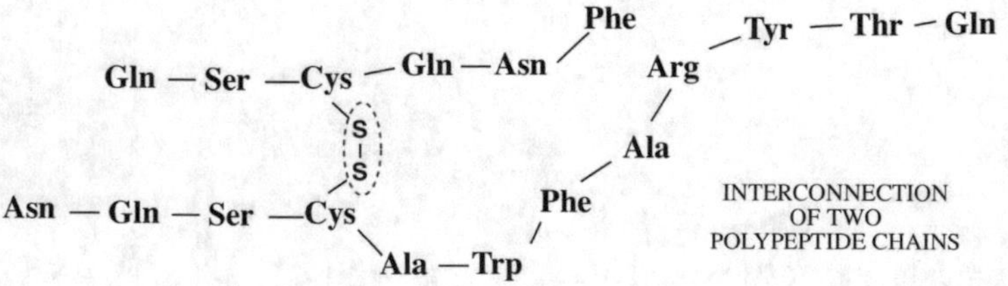

INTERCONNECTION
OF TWO
POLYPEPTIDE CHAINS

In the same way the long polypeptide chain can also form a loop as it is shown below:

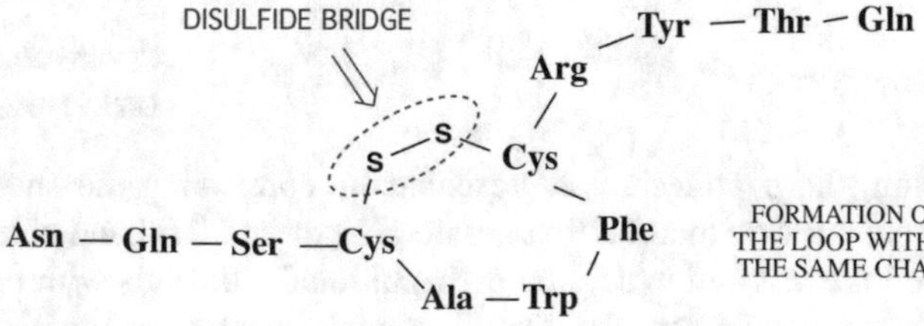

FORMATION OF
THE LOOP WITHIN
THE SAME CHAIN

We have argued that the peptide bond is essentially related to the amide functional group. The resonance structures of amides which we have discussed earlier can also be used in the representation of the electronic structure of polypeptide molecules.

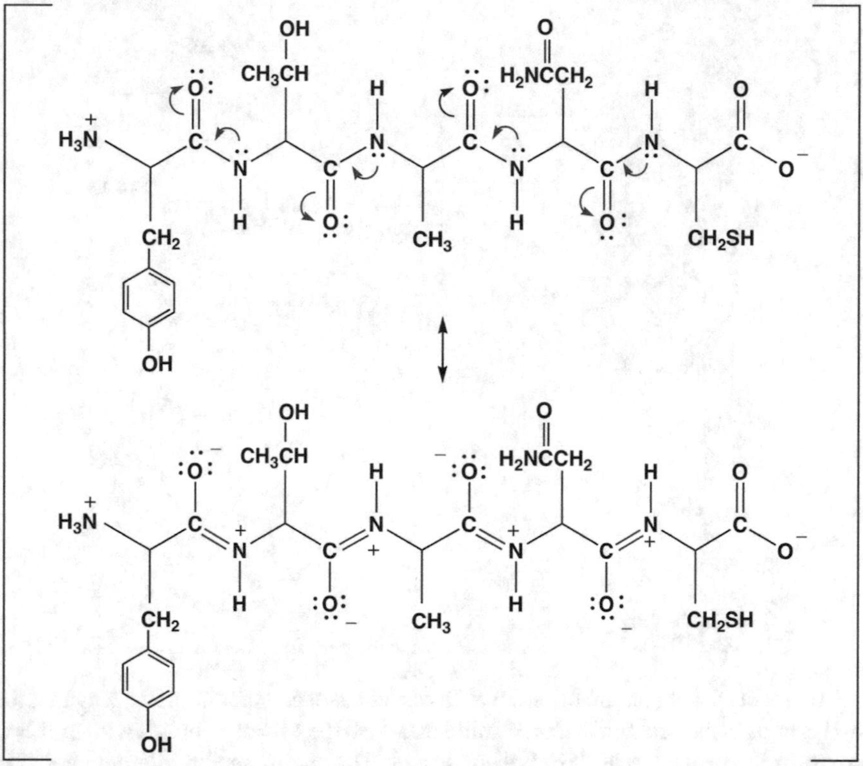

Inspection of these structures makes it clear that the chemical bonds between carbon and nitrogen have double bond character. Consequently, the rotation around these bonds is restricted as is the rotation around the double bonds in alkenes. This restricted rotation makes polypeptide chains rigid. Because of their structural rigidity, polypeptide chains appear in two forms: **α-helix** and **β-sheet**.

Each complete turn of the helix requires 3.6 amino acids (residues) in the chain which contains N–C–C–N bonds. The helix is additionally stabilized by hydrogen bonds between the CO group of residue n and the NH group of residue n + 4, i.e. between spatially close amino acids (see the next diagram).

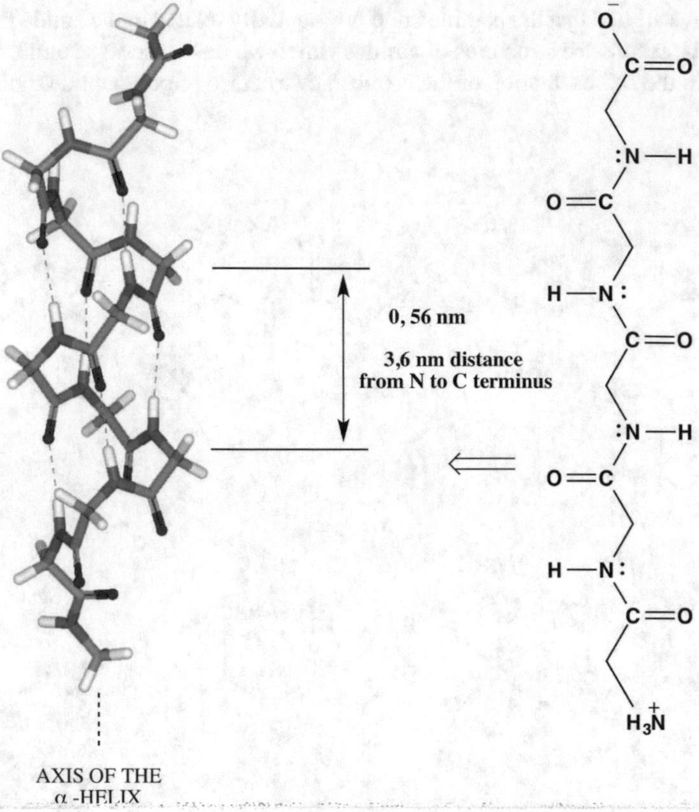

0, 56 nm

3,6 nm distance
from N to C terminus

AXIS OF THE
α -HELIX

In β-sheets the chains are oriented with respect to each other in such a way that the N–H groups in the backbone of one strand establish the hydrogen bonds with the C=O groups in the backbone of the adjacent strand. The chains can be interconnected in two ways: **antiparallel** or **parallel**. In the antiparallel arrangement, as represented in the scheme below, the successive β strands alternate directions so that the N-terminus of one strand is adjacent to the C-terminus of the other strand.

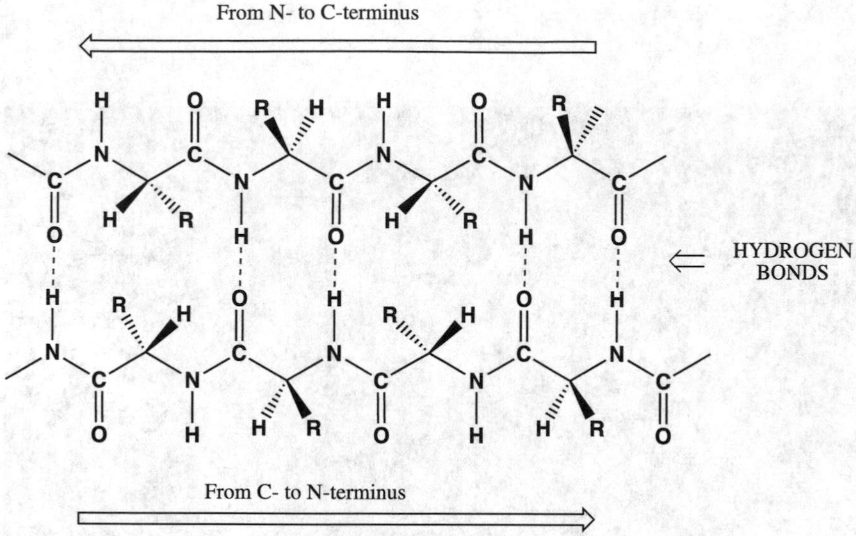

From N- to C-terminus

HYDROGEN BONDS

From C- to N-terminus

β - SHEET

In the parallel structure shown below the N-termini in both strands are oriented in the same direction as are the C-termini.

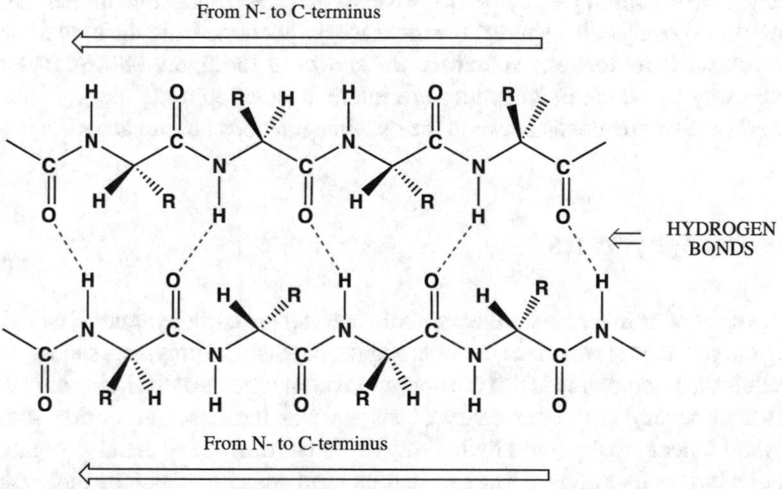

From N- to C-terminus

HYDROGEN BONDS

From N- to C-terminus

β - SHEET

Since the **protein molecules** consist of polypeptide chains interconnected by disulfide bonds their structure can be represented by simple graphical models in which the α-helices and β-sheets are depicted as in the figure below.

β - SHEETS   α - HELIX

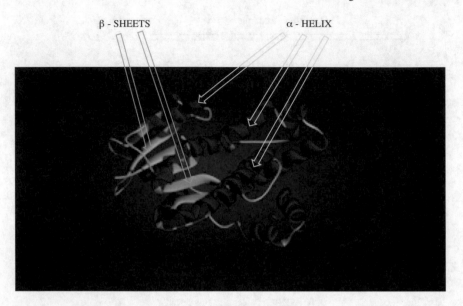

As we have seen from the overview of chemistry of polypeptides three levels of the structure of proteins can be discerned. The first level is the **primary structure** of proteins (polypeptides) which is simply the linear order of amino acids in the chain starting from the N-terminus up to the C-terminus. Formation of α-helices and β-sheets is the **secondary structure of proteins** and it is responsible for the shape of the polypeptide molecule. Aggregation of these polypeptides into the final molecule of the protein is its **tertiary structure** (as shown in the figure above). The most common way by which polypeptides are interconnected into the tertiary structure includes the disulfide bonds between the cysteine residues (amino acids).

11.2 Carbohydrates

Another class of primary metabolites soluble in water are carbohydrates. To this class belong the substances which we know as sugars, but also the polymers such as starch and cellulose, the compounds that are ubiquitous components of the living organisms. The first discovery in the chemistry of sugars was the observation that some of them yield other carbohydrates by hydrolysis. In contrast, there are also sugars that cannot be further hydrolyzed. They are called **monosaccharides**. The molecules of monosaccharides can be joined into the larger molecules called **oligosaccharides** or even **polysaccharides**.

The name "carbohydrates" comes from their composition which is represented by the general formula $C_m(H_2O)_n$ which resembles the combination of carbon and water. Such composition implies that the molecule of monosaccharide consists of a hydrocarbon chain with attached hydroxyl groups. More detailed analysis shows that most of the monosaccharide molecules have hydrocarbon chains which are from five to seven carbon atoms long. From the study of chemical behavior of different monosaccharides, it follows that some of them exhibit reactions typical for aldehydes and others show reactions typical for ketones. The monosaccharides can therefore be classified into two groups: **aldoses** and **ketoses**. In their nomenclature, the names of carbohydrates are formed from the root based on the number of C-atoms, the fundamental functional group (ketone or aldehyde) and the suffix **–ose**. While the aldehyde group always contains the first carbon atom of the chain, keto-group in all known carbohydrates appears at the second carbon atom of the chain. Some monosaccharides named in accordance with these rules are represented in the scheme below.

Carbon atoms not situated at the ends of the molecule always have four different substituents and thus represent the stereogenic centers. Such chiral C-atoms are labeled with asterisks in the scheme above. Since every chiral carbon atom can have two configurations, every monosaccharide can have 2^n stereoisomers with n being the number of stereogenic centers. For instance, aldohexose can have 16 stereoisomers. This is a good example of a large variety of possible structures for organic compounds. As we recall from the chemistry of amino acids, the nature is very selective in generation of specific isomers. The most common monosaccharide found in living organisms is the aldohexose traditionally called **D-glucose.**

The configurations of all chiral C-atoms in the glucose molecule are shown in the formulas of the scheme below. While the structures A and B represent glucose molecule using wedge-dash notation, C is the Fischer projection formula and D is the shorthand Fischer formula. The horizontal lines label the positions of −OH groups. Remember that the substituents written on the left and right side of the vertical line are located above the plane of drawing.

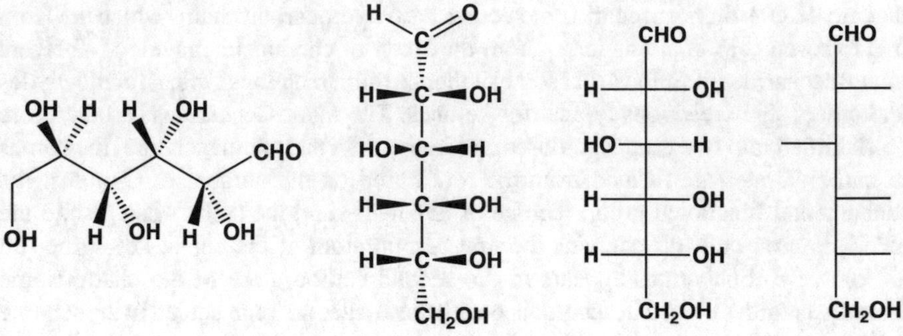

A   B   C   D

D-glukoza

Let us give some more examples of monosacchaides. **D-manose** belongs to aldo-hexoses, **D-ribose** and **D-arabinose** belong to aldopentoses and **D-fructose** is the most important ketohexose. The label D represents the relative configuration in accordance with the rules which were discussed in the section on amino acids. Since the monosaccharide molecule possesses several stereogenic C-atoms the labels D or L are used only for the chiral carbon that is furthest from the aldehyde or the ketone group. As in case of amino acids which all have L-configuration, all the natural monosaccharides have exclusively the same **D-configuration**. This is another example of selectivity in nature!

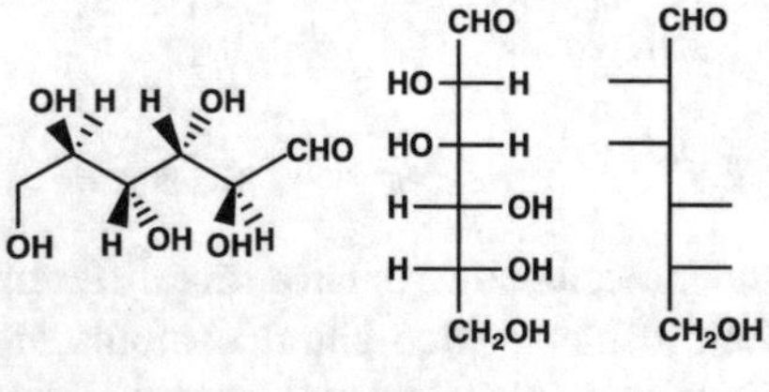

D-manoza 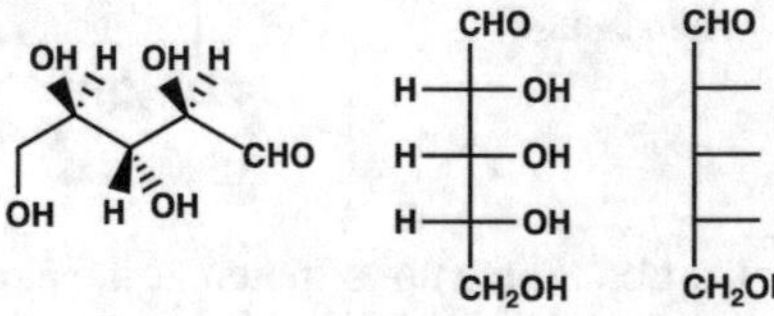   **D-riboza**

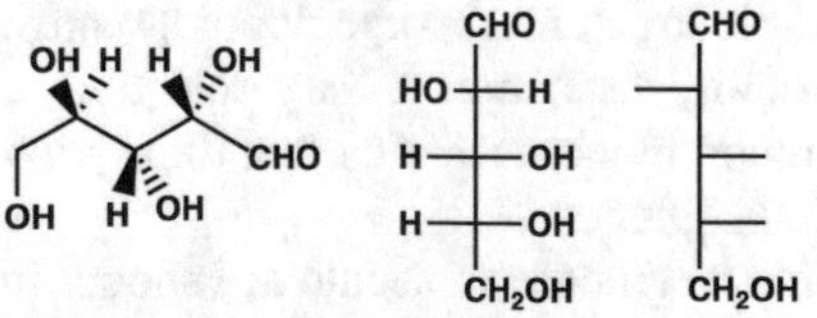

D-arabinoza 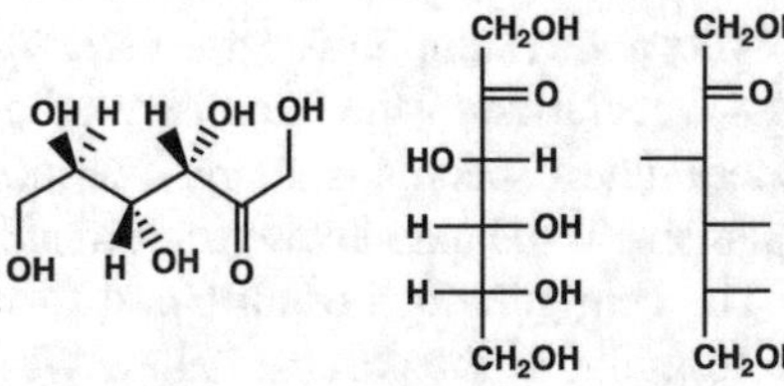   **D-fruktoza**

The evidence that in natural products, especially in primary metabolites, some structural patterns are repeated in different structures implies that most organisms have common biosynthetic pathways. At the end of XIX century when the biosynthetic pathways were not known, **Heinrich Kiliani** and **Hermann Emil Fischer** have developed a simple model which could explain how stereochemically pure long-chained monosaccharides can be formed from the simple molecules. The simplest, parent monosaccharide is **aldotriose** with only one chiral carbon atom. Hence, there are D- and L-aldotrioses. Since aldotriose is the product of mild oxidation of glycerol it can also be called **glyceraldehyde**.

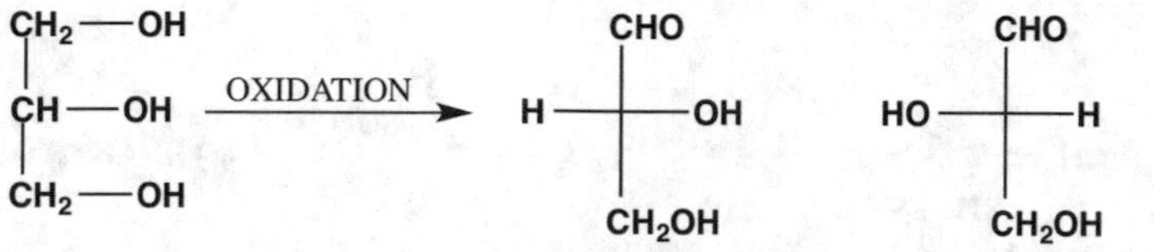

| Glycerol | D-Glyceraldehyde (D-Aldotriose) | L-Glyceraldehyde (L-Aldotriose) |

To extend the molecule of aldotriose (glyceraldehyde) by adding an extra carbon atom, Kiliani and Fischer applied the already known reaction of nucleophilic addition to the aldehyde group. They have used the cyanide ion ^-CN as the nucleophile. After the addition of cyanide to the aldehyde group whose carbon atom is achiral, the said carbon atom becomes chiral in the product because of the four different substituents attached. This change in stereochemistry is shown in blue in the following scheme. The products are two diastereomers in which the new stereogenic center (the C-atom bound to the CN group) can have D- or L-configuration.

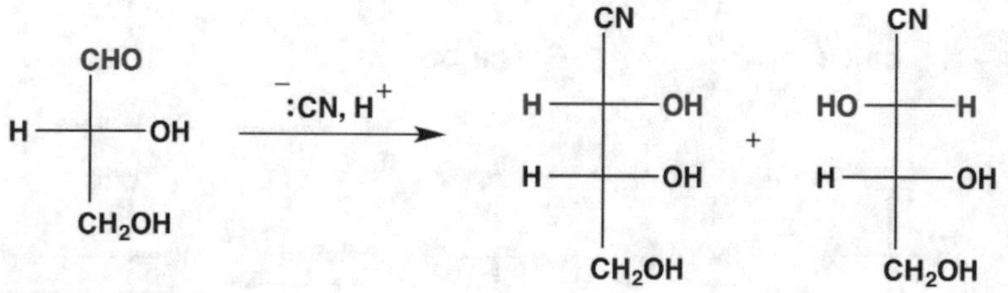

D-Glyceraldehyde
(D-Aldotriose)

Since the aldehyde carbon is bound to oxygen by the double bond this carbon is coplanar with all three substituents. The ^-CN nucleophile can then attack this carbonyl plane from either side with equal probability which leads to the formation of two products that differ in their relative configurations, D or L (scheme below).

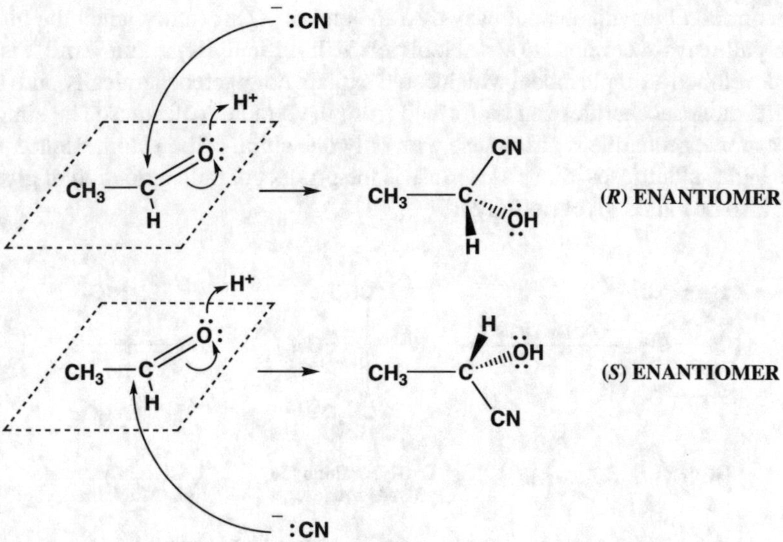

If the cyanhydrine products are then first oxidized to acids and subsequently reduced to aldehydes, the final products are two new monosaccharides, the aldotetroses **D-erithrose** and **D-threose**:

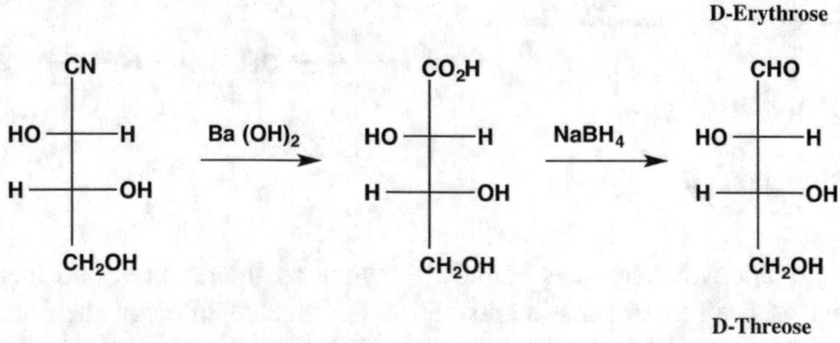

This **Kiliani-Fischer synthesis** can be continued starting from these new tetroses (D-erithrose and D-threose) to prepare four new monosaccharides (aldopentoses), because every tetrose can yield two new pentose diastereomers. By repeating this series of reactions, it is possible to extend the carbon chain further to get hexoses. However, if the synthesis starts with D-glyceraldehyde all the product molecules of corresponding aldoses will have D-configuration on the C-atom which is furthest from the aldehyde group. These monosaccharides belong to the **D-series**. On the other hand, aldoses prepared from L-glyceraldehyde will have the last C-atom with L-configuration. These monosaccharides belong to the **L-series**. All living organisms produce only the D-series of carbohydrates!

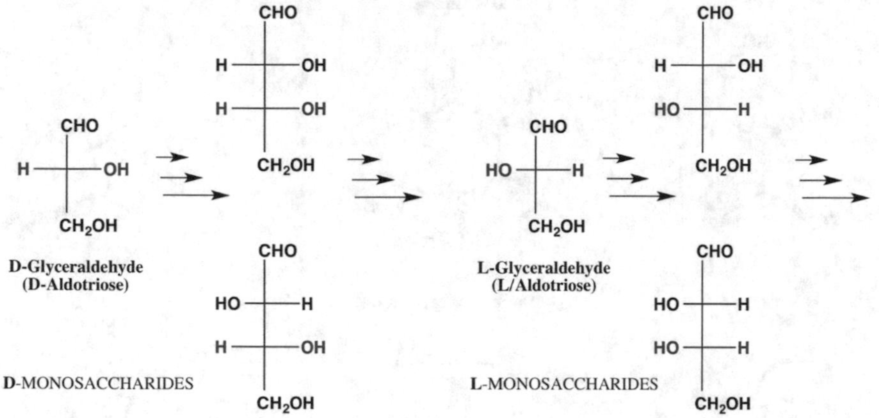

11.2.1 *Cyclic Structures of Monosaccharides*

The representation of monosaccharide molecules as open-chain structures is only partially correct. In solution, the molecules with five- and six carbon atoms can also appear in cyclic forms which are in equilibrium with the open-chain structures. The cyclization process represents the intramolecular nucleophilic attack of the hydroxyl group on the aldehyde or keto group.

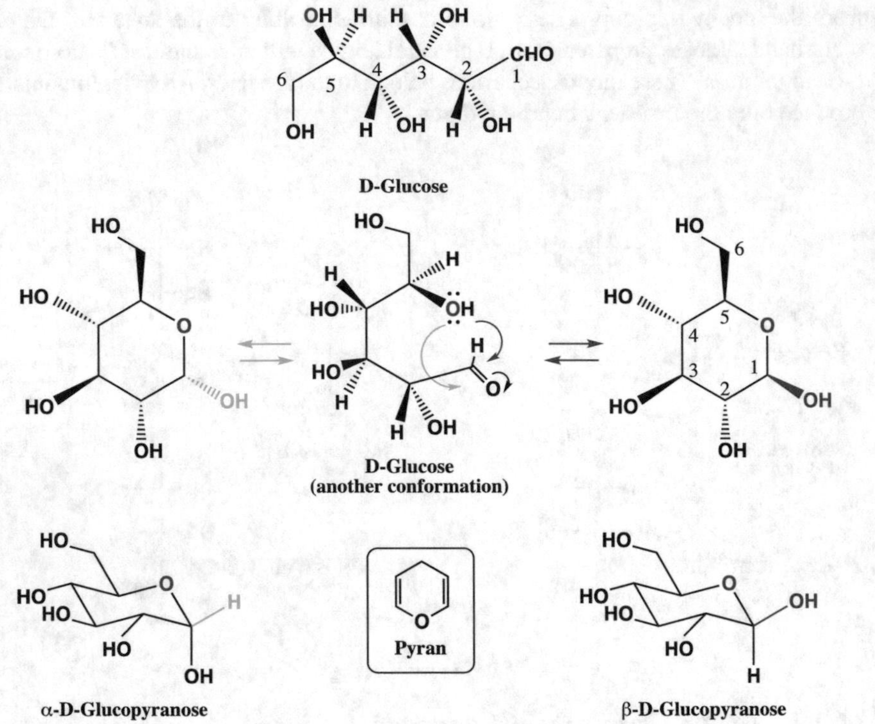

D-Glucose

D-Glucose
(another conformation)

α-D-Glucopyranose

Pyran

β-D-Glucopyranose

The carbon atom which in the open-chain form was part of the carbonyl group after cyclization becomes chiral and is called the **anomeric carbon atom**. In the cyclic form the anomeric carbon can have two configurations which are labeled with Greek letters α and β. Stereoisomers that differ in the configuration at this new chiral center are called **anomers**. For example, the cyclic form of D-glucose can have α and β anomers named **α-D-glucopyranose**, and **β-D-glucopyranose**. The word **pyranose** has been taken from the analogous cyclic ether with the six-membered ring which is known as **pyran**. In the first chapter of this book, we have claimed that the substituted cyclohexane ring is most stable when its substituents are in equatorial positions. The anomeric carbon atom is an exception; the most stable anomer has the OH group in the axial position which makes it the α-anomer. This is called the anomeric effect.

Anomers also appear in cyclic form of the five-membered rings, for instance in monosaccharides fructose and ribose. These cyclic molecules are named **furanoses** because of their similarities to the cyclic ether **furan**. Starting from D-fructose and D-ribose, the cyclization yields anomers α- and **β-D-fructofuranose**, as well as α- and **β-Dribofuranose**, respectively.

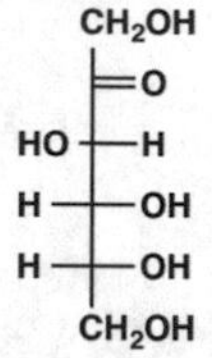

D-Froctose 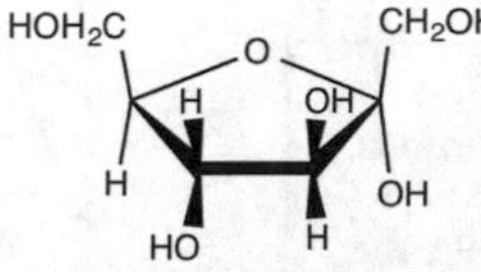 **α-D-Fructofuranose** 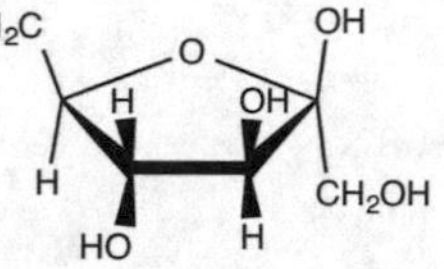 **β-D-Fructofuranose**

Furan

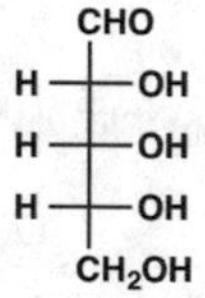

D-Ribose 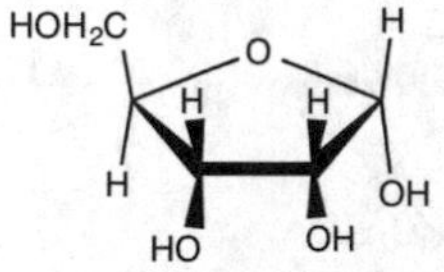 **α-D-Ribofuranose** 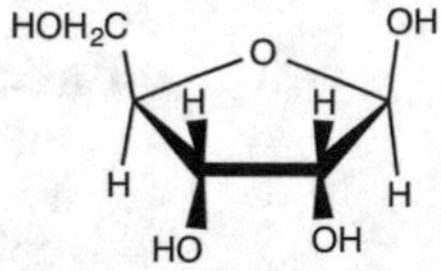 **β-D-Ribofuranose**

In analogy with pyranoses, the furanoses are also in equilibrium with the open-chain forms. The new functionality which appears in the cyclic forms is **hemiacetal**, the structure already mentioned in our discussions on alcohol addition to the carbonyl group. The hemiacetals can easily undergo hydrolysis yielding the open chain monosaccharide molecule.

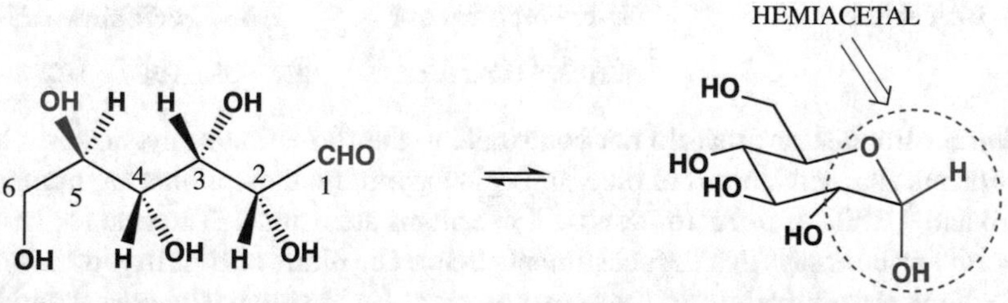

D-Glucose **α-D-Gulucopyranose**

Representation of cyclic forms could be simplified by using special shorthand notation called Haworth formula (in the honor of **Sir Norman Haworth**). Fischer formulas can be easily transformed into Haworth formulas by the rules given in the following scheme.

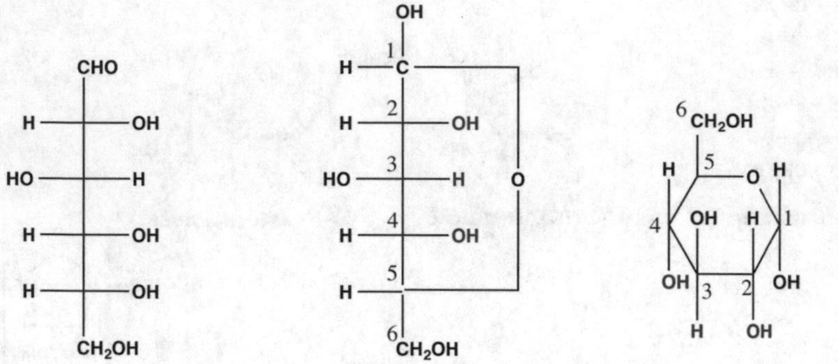

D-Glucose   **D-Glucopyranose**   **α-D-Glucopyranose**

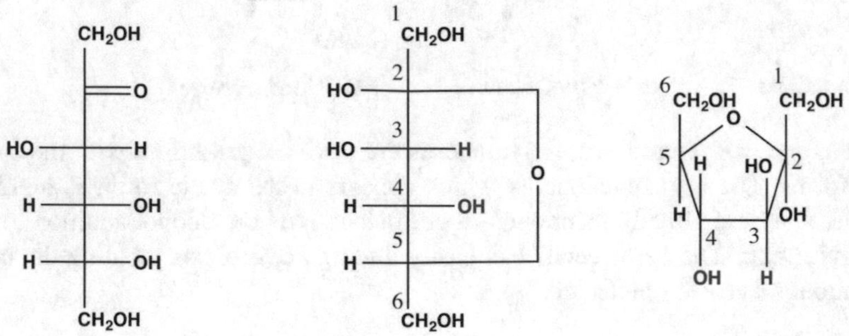

D-Fructose   **D-Fructofuranose**   **α-D-Fructofuranose**

FISCHER`S FORMULA        HAWORTH`S FORMULA

Chiral carbon atoms that do not contribute to the formation of cyclic form have substituents attached labeled in blue. In D-glucopyranose these atoms are numbered as 2, 3 and 4 while in D-fructofuranose these atoms are 3 and 4. The rule for writing Haworth formulas says that the substituents **below the plane of the ring in Haworth projections are equivalent to those on the right-hand side of the corresponding Fischer projection** (shown in blue).

11.2.2   Disaccharides and Polysaccharides

The hydroxyl group on the anomeric carbon of the monosaccharide molecule can be replaced by another monosaccharide molecule. The resulting dimer is called **disaccharide**. Binding of α-D to β-D-fructofuranose gives disaccharide α-D-glucopyranosyl-β-D-fructofuranoside, the disaccharide known under the name **sucrose** which is commonly used in food preparation (see the scheme). Analogously, β-D-galactopyranose can bind with β-D-glucopyranose into disaccharide β-D-galactopyranosyl-β-D-glucopyranoside or **lactose**, the main sugar component of milk. The chemical bond by which the monosaccharide units are joined is called the **glycosidic bond**. In lactose, the carbon atom 1 on one monosaccharide is bound to the carbon 4 of the second monosaccharide molecule, and such bond is known as the **1,4 glycosidic bond**. This is the most frequent mode of bonding between monosaccharide units.

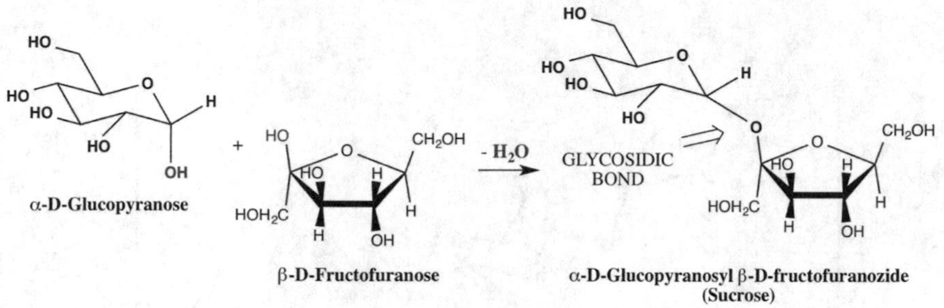

α-D-Glucopyranose

β-D-Fructofuranose

α-D-Glucopyranosyl β-D-fructofuranozide
(Sucrose)

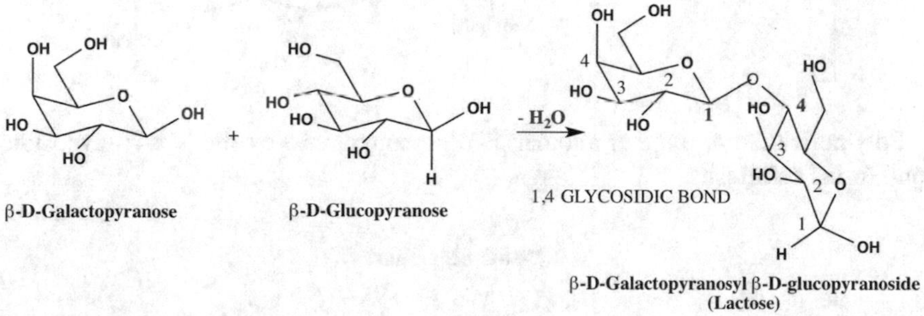

GLYCOSIDIC BOND

β-D-Galactopyranose

β-D-Glucopyranose

1,4 GLYCOSIDIC BOND

β-D-Galactopyranosyl β-D-glucopyranoside
(Lactose)

In the same way, the glucose molecules can be connected by glycosidic bond into long polymeric structures of **polysaccharides**. The structure of the polysaccharide formed from glucose depends mostly on the starting anomer. Polymerization of α-D-glucopyranose by the α(1,4)-glycosidic bond yields **amylose**, the polysaccharide which is component of **starch**.

α-D-Glucopyranose

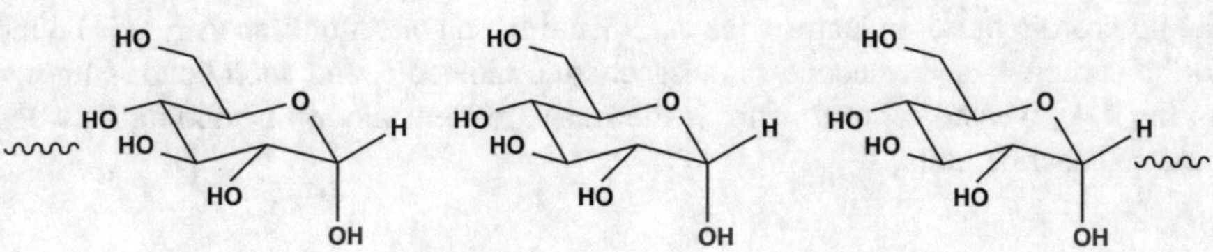

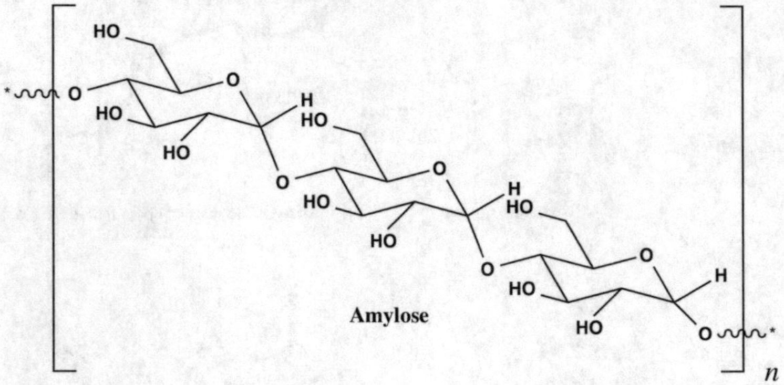

Amylose

Polymerization of another anomer, β-D-glucopyranose by the β(1,4)-glycosidic bond forms **cellulose**.

B-D-Glucopyranose

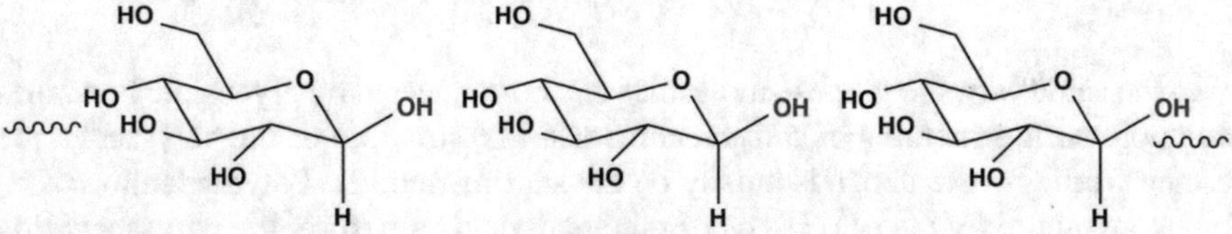

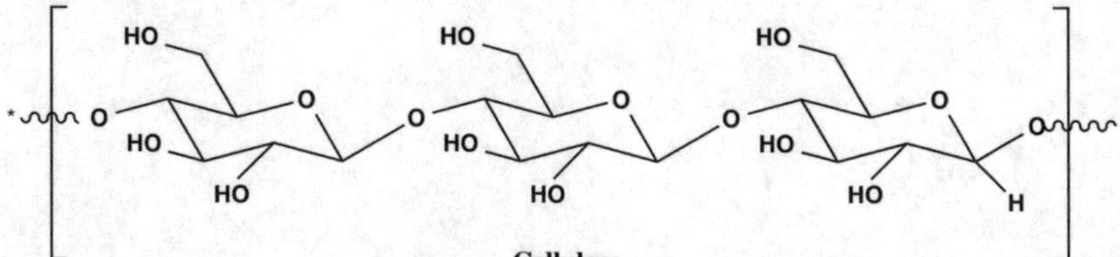

Cellulose

11.3 Glycosides and Nucleotides

Glycosides are compounds in which the substituent is bound to the anomeric carbon. This group comprises numerous natural products; the primary metabolites. The most important are **nucleosides**, the basic structural units from which **nucleotides** and **nucleic acids** are formed. The monosaccharide component of nucleoside is **D-ribose** or its partially reduced form **D-deoxyribose**. In biochemistry and molecular biology, the most important are the nucleosides in which ribose is bonded to the heterocyclic molecules called **nucleic bases**. The most common nucleic bases are **pyrimidines** and **purines**.

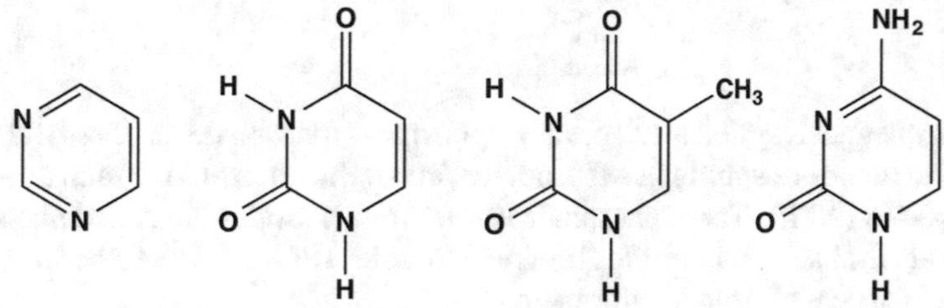

| Pyrimidine | Uracil | Thymine | Cytosine |

PYRIMIDINE BASES

Pyrimidines are derivatives of the cyclic amine **pyrimidine** while purines are derived from the bicyclic amine **purine**.

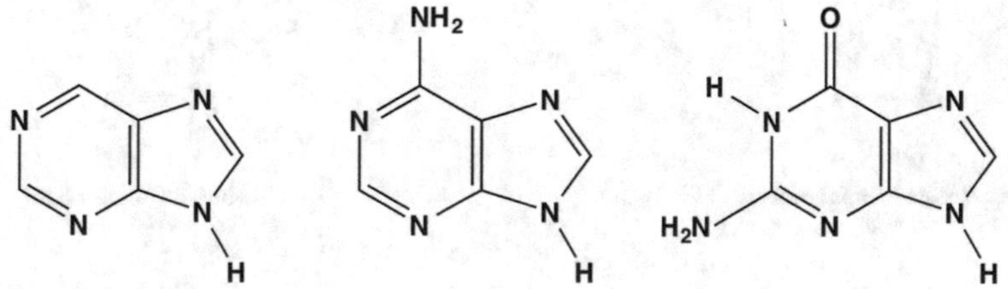

| Purine | Adenine | Guanine |

PURINE BASES

The glycoside that consists of adenine and **D**-ribose (in its cyclic form ribofuranose) is **adenosine**.

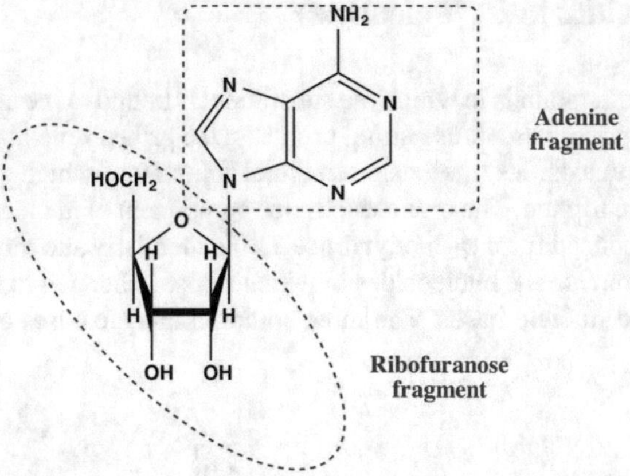

Adenosine

In biochemical systems adenosine appears as phosphate in three forms, as **adenosine monophosphate (AMP)**, **adenosine diphosphate (ADP)** and **adenosine triphosphate (ATP)**. These phosphate esters of nucleosides are called nucleotides and they are the basic building blocks of nucleic acids **DNA** and **RNA**, the compounds which are carriers of genetic information.

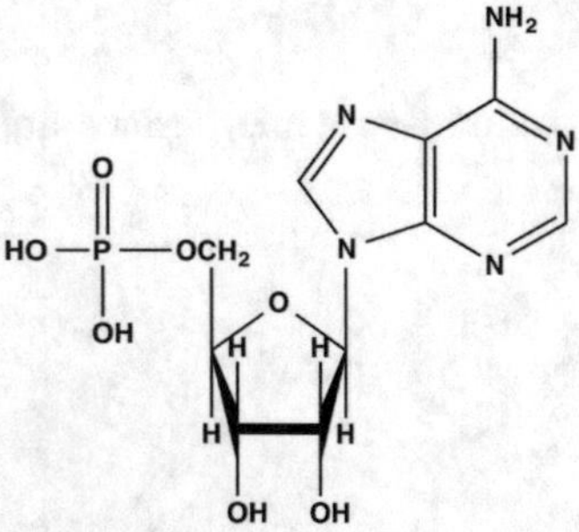

Adenosine monophosphate
AMP

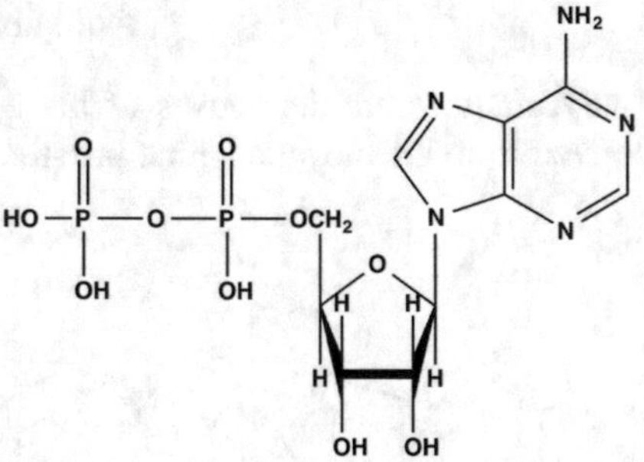

Adenosine diphosphate
ADP

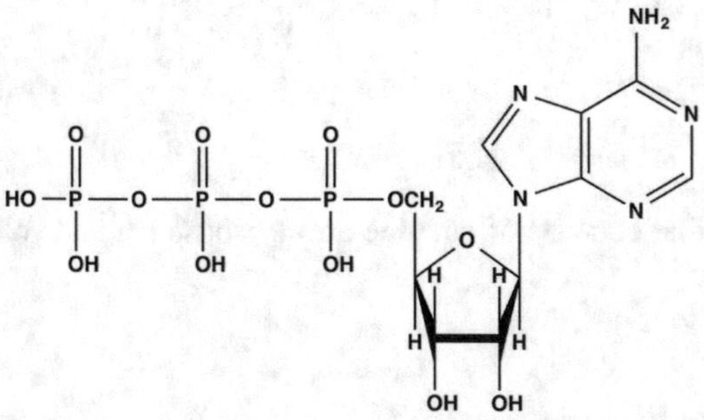

Adenosine triphosphate
ATP

In the enzyme catalyzed reaction ATP can be hydrolyzed by losing one phosphate group PO_4^{3-}, with the release of large amount of energy that has been accumulated in the P-O bond. This reaction serves as the source of energy for the processes in living cells.

Nucleic acids are polymers made from nucleotides. In living organisms two main sorts of nucleic acids are present: **ribonucleic acid**, **RNA** and the **deoxyribonucleic acid**, **RNA**. The building blocks of nucleic acid molecules comprise five nucleotides bound either to the **ribofuranose** (as in RNA) or to the **deoxyribofuranose** (as in DNA) and to the phosphate group. Nucleotides which are phosphates of adenine, thymine, guanine and cytosine are the components of DNA. In RNA the thymine unit is replaced with uracil. Polymeric molecule of nucleic acids can exist as either **single-stranded** or **double-stranded** chains. In the single-stranded chain the ribose molecules are interconnected via the phosphate groups. The structure of the nucleic acid chain is represented in the scheme below. RNA and DNA differ in the substituent Z on C_2 carbon atom of the monosaccharide unit. If Z is the OH group, the fragment of the nucleic acid molecule would represent RNA, but if Z is the hydrogen atom the structure would represent DNA. Ribose and deoxyribose differ in this detail only.

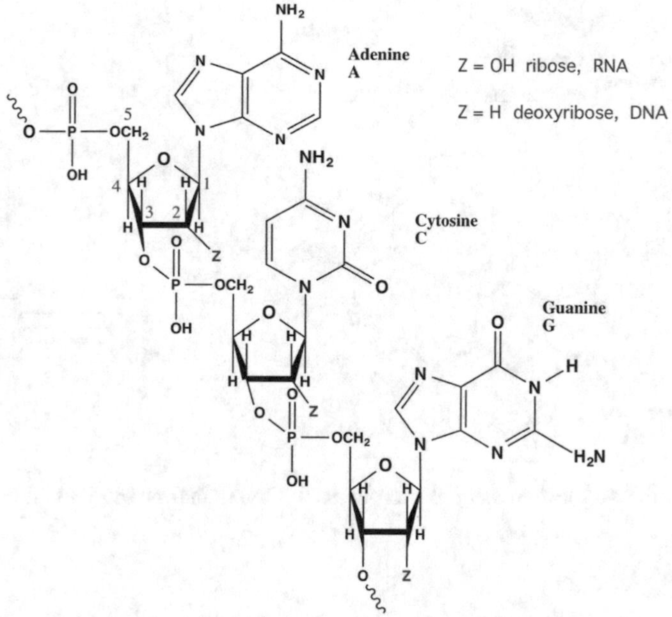

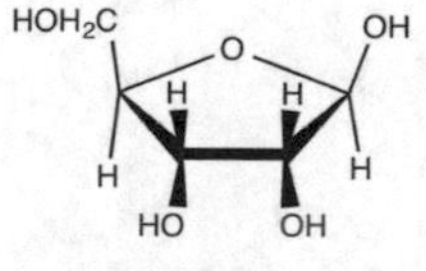

β-D-Ribofuranose
Ribose

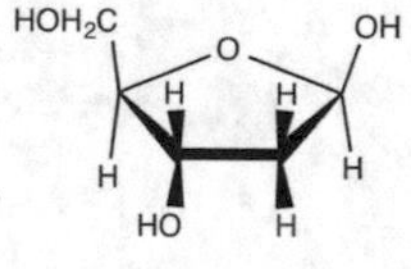

β-D-Deoxyribofuranose
Deoxyribose

The double-stranded DNA consists of two parallel chains folded into α-helix. The chains are joined by the hydrogen bonds that follow the **rules of complementarity**. In accordance with these rules, adenine is always bound with thymine of the other chain and guanine is always bound to cytosine (see the following scheme). Principles of molecular complementarity do not apply only to nucleic acids. Rather, they represent general rules for the formation of various **supramolecular structures**. More about the supramolecular structures will be discussed in following chapter. If nucleotides are labeled by their first letters, the complementary pairs would be AT and GC, respectively. The structural key for complementarity is the number of hydrogen bonds. There are two such bonds in the AT pair and three in the GC pair. For example, it is impossible for G to be bound to T: the nucleotides are **selective**.

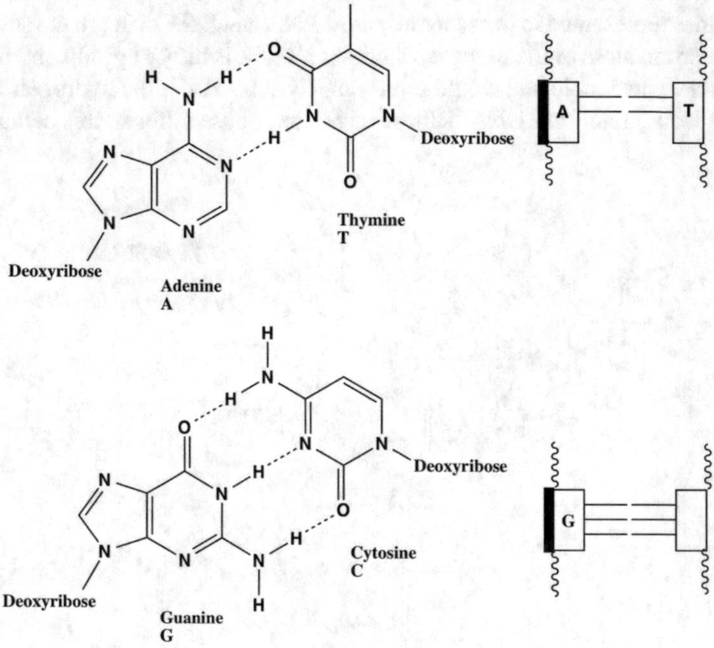

Based on complementarity principle two DNA chains bind as shown in the scheme:

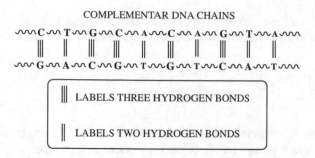

In biochemical processes the DNA molecule can be separated in two single-stranded molecules. One of such single-stranded intermediates serves as a matrix on which the complementary chain can be synthesized. This is the main principle of DNA **replication**. The mechanism and numerous details of this process are studied in biochemistry.

11.4   Lipids

At the beginning of this Chapter, we have mentioned that one of the oldest methods for isolation of organic natural products is extraction. Primary metabolites discussed up to now were mainly ionic or polar compounds soluble in water or in polar solvents. Organic natural products that belong to different structural classes but which have in common the solubility in nonpolar solvents, are called **lipids**. Some of the lipids are esters of long-chained fatty acids and they can be readily hydrolyzed under acidic or basic conditions. In this group we encounter **waxes**, **fats** and **phospholipids**.

11.4.1   Waxes

Waxes are structurally the simplest lipids being esters of long-chained carboxylic acids and long-chained alcohols. The chains can have 12 or more carbon atoms. The ester represented in the following scheme is the main component of wax produced by bees. It can be easily saponified into the corresponding long-chained acid and alcohol.

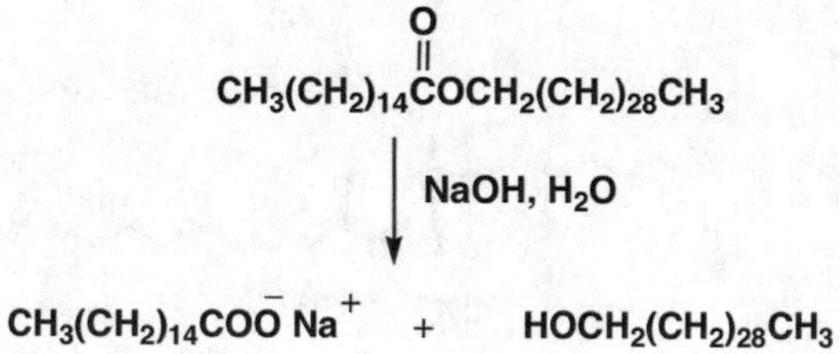

$$CH_3(CH_2)_{14}\overset{\displaystyle O}{\overset{\|}{C}}OCH_2(CH_2)_{28}CH_3$$

$$\downarrow \quad NaOH, H_2O$$

$$CH_3(CH_2)_{14}COO^- Na^+ \quad + \quad HOCH_2(CH_2)_{28}CH_3$$

The reaction of saponification has already been discussed in the chapter on esters. This reaction is the basic procedure for decomposition of ester lipids.

11.4.2 Fats

Fats are esters of long-chained fatty acids and the alcohol glycerol. Since glycerol has three hydroxyl groups its esterification requires three molecules of fatty acids. The chains in the molecules of fatty acids can be saturated or partially non-saturated. Some examples are given in the following table.

$CH_3(CH_2)_{10}COOH$	**Lauric acid**
$CH_3(CH_2)_{12}COOH$	**Myristic acid**
$CH_3(CH_2)_{14}COOH$	**Palmitic acid**
$CH_3(CH_2)_{16}COOH$	**Stearic acid**
$CH_3(CH_2)_7CH=CH(CH_2)_7COOH$	**Oleic acid**
$CH_3CH_2CH=CHCH_2CH=CHCH_2CH=CH(CH_2)_7COOH$	**Linolenic acid**

Fats and oils differ according to the structures of their fatty acid. While fats are made of saturated acids, the oils are esters of unsaturated fatty acids. As we know from the chemistry of hydrocarbons, unsaturated compounds such as alkenes or alkynes can be saturated by addition of hydrogen in a metal-catalyzed reaction. Using the same method, natural oils can be transformed to fats that are frequently called synthetic fats. The process of such hydrogenation which yields the fat called tristearin is represented in the following scheme.

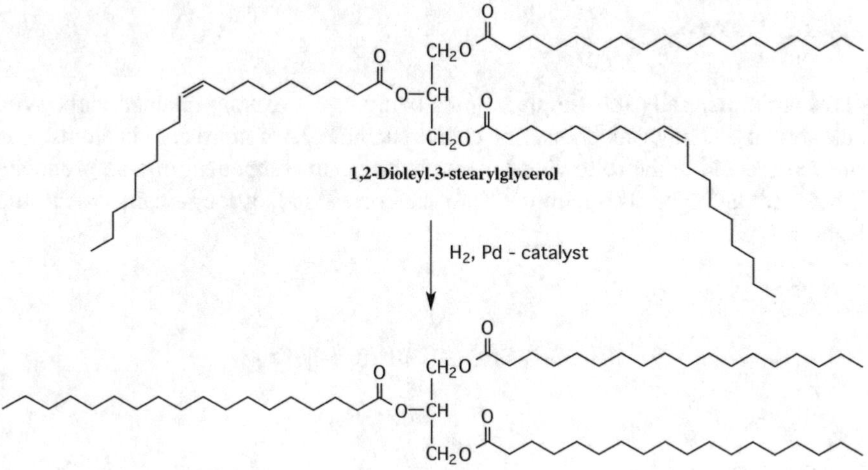

1,2-Dioleyl-3-stearylglycerol

H_2, Pd - catalyst

**Tristearyl glycerol
(Tristearine)**

Because of the presence of double bonds, the digestion of oils is easier than the digestion of saturated fats. Recent nutritional recommendation is to use the oils that consist of the **ω-3 fatty acids**. Another name for them is ***n*-3 fatty acids**. These compounds have hydrocarbon chains consisting of 18–22 carbon atoms in which the first of several double bonds resides on the third C-atom counting from the alkyl (not carboxyl!) terminus of the molecule. Linoleic acid is the representative example of ω-6 fatty acids. In addition, the molecules of unsaturated fatty acids have Z-configuration around the carbon–carbon double bonds!

Saponification of fats with KOH or NaOH yields **soaps**, the sodium or potassium salts of fatty acids and glycerol. The behavior of soaps in water solutions and the formation of micelles have already been discussed.

Tristearylglycerol
(Tristearin)

↓ **KOH**

Potassium stearate
(soap)

Let us mention the old tradition known from the history of ethnology for the preparation of soaps by boiling animal fat with the ash that remains after burning of the beech wood. This ash contains potassium carbonate and hydroxide which serve as bases for the saponification. Interestingly, in the Arabic language the word for ash is *kali*. The terms *alkali* or *kalium* (for the element potassium) were derived from this root. Arabic culture and tradition strongly influenced the European alchemy and its transformation into modern chemistry.

11.4.3 Phospholipids

The property of the glycerol-based lipids to form micelles even without saponification is especially pronounced in phospholipids. Basic structural unit of the phospholipid molecule is **phosphatidic acid** shown in the scheme below. The substituents R and R´ are long-chained alkane or alkene groups. The phosphate group can be combined with different structures such as **choline** or **serine**, the molecules which play important roles in metabolism.

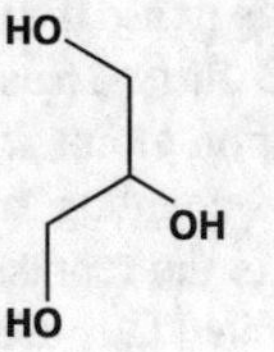

Glycerol 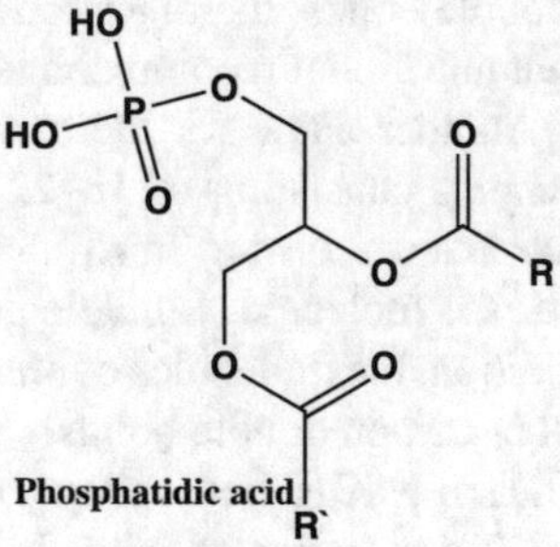   Phosphatidic acid   Choline   Serine

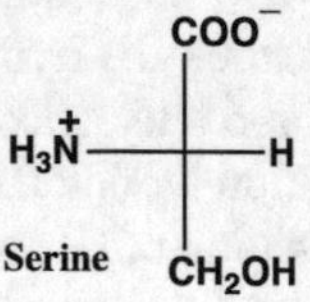

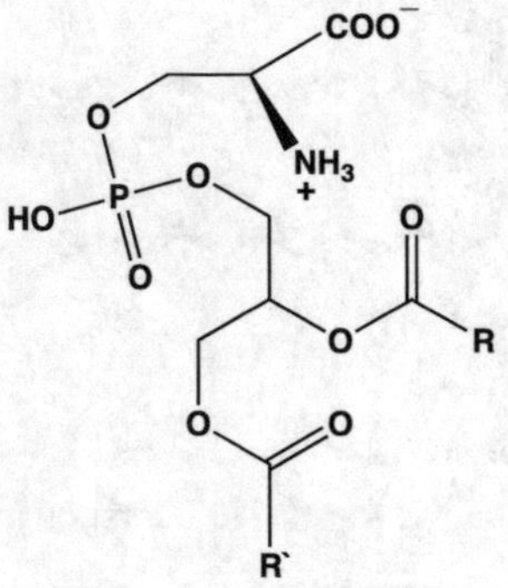

Phosphatidylserine 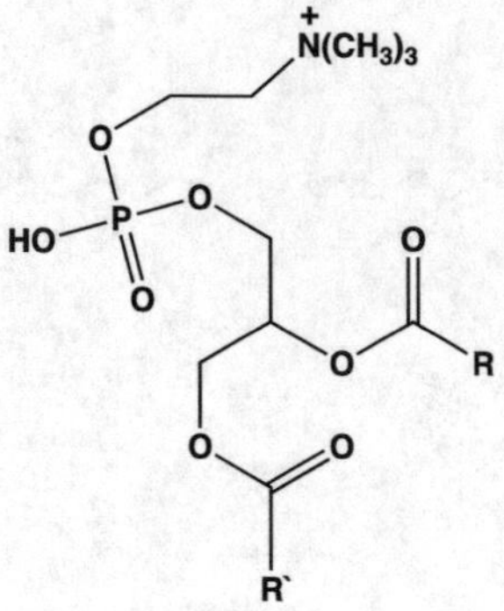   Phosphatidylcoline
                                              (Lecithin)

The compounds similar to **lecithin** are the components of cellular membranes because they can form double-layered structures in which the hydrophobic sides are oriented towards each other. Being hydrophilic the choline group is oriented towards the water solvent.

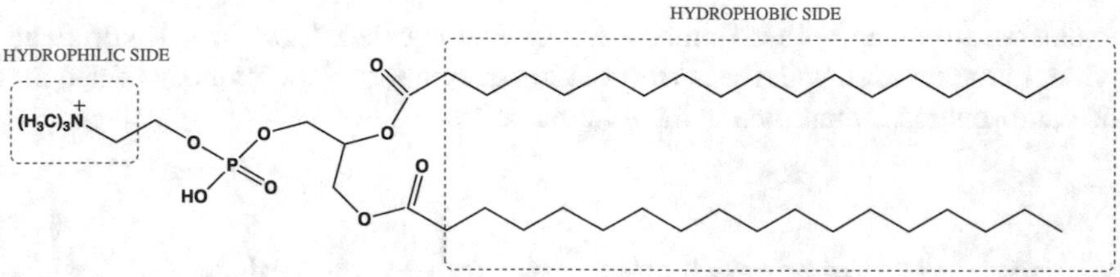

HYDROPHOBIC SIDE

HYDROPHILIC SIDE

$(H_3C)_3\overset{+}{N}$

HO

1,2-Distearylphosphatidyl choline

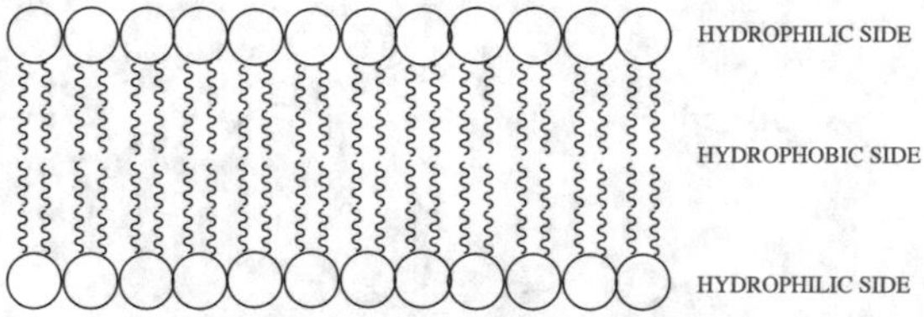

HYDROPHILIC SIDE

HYDROPHOBIC SIDE

HYDROPHILIC SIDE

DOUBLE-LAYER STRUCTURE OF CELLULAR MEMBRANE

Such membranes prevent the passage of Na^+ and K^+ ions, the process important for the transmission of electrical signals between living cells. Being hydrophilic, these ions cannot pass through the hydrophobic layer of the membrane. For ion transport through membranes, the cells have developed special catalytic systems which require specific groups of enzymes.

11.4.4 Terpenes and Steroids

In contrast to fats and phospholipids terpenes and steroids cannot be hydrolyzed to simpler units. However, **Otto Wallach** in XIX century and **Leopold Ružička** (Croatian Nobel laureate) in the first half of XX century have discovered that these compounds can be decomposed into simpler structures that consist of five carbon atoms. These units are structural analogs of the alkene named **isoprene** (2-methylbuta-1,3-diene). Isoprene can be regarded as the molecule containing the head and the tail.

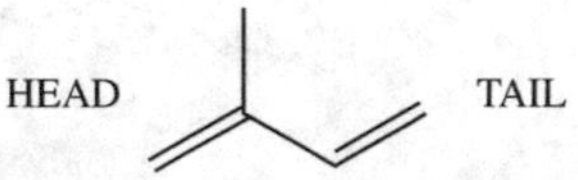

HEAD   TAIL

Isoprene

Isoprenic units can be interconnected in different ways: head-to-head, tail-to-tail, head-to-tail or tail-to-head. Oligomers of such isoprene units are called **terpenes**. The molecule with two isoprenes is named **monoterpene**, the molecule with four units is **diterpene**. Three isoprene monomers form **sesquiterpenes**. One of the most important terpenes is **squalene**, the primary metabolite that is present in all living organisms. In contact with the atmospheric oxygen squalene is oxidized to **squalene-oxide**. In the reaction with the corresponding enzyme squalene-oxide is folded into the conformation represented in the scheme bellow.

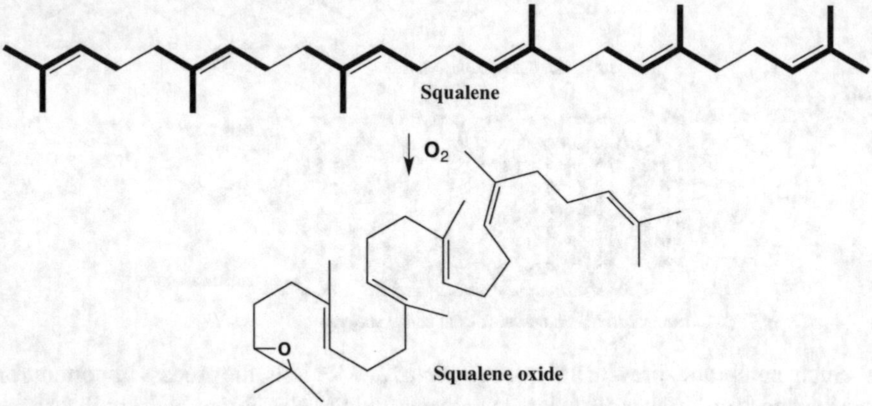

Squalene

O_2

Squalene oxide

In the subsequent reaction step the epoxide ring is protonated to carbocation which activates successive cyclizations of four double bonds to give the tetracyclic structure of **lanosterol**.

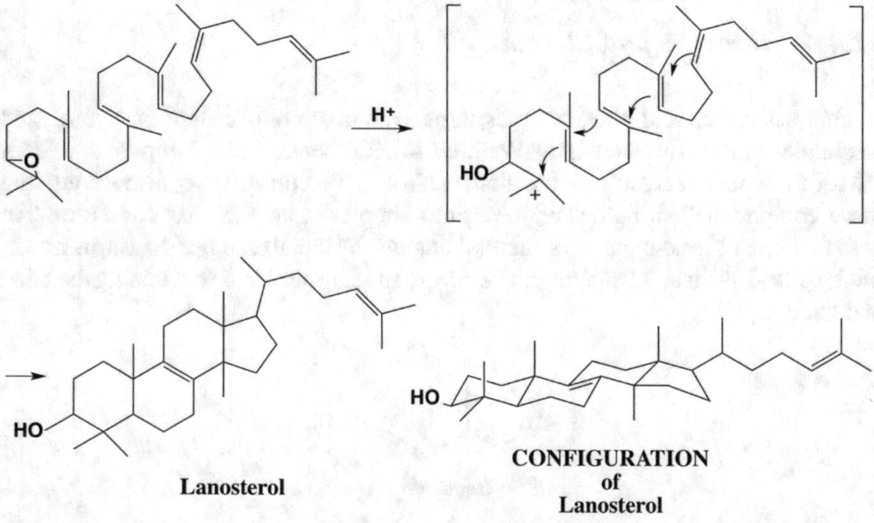

H+

HO

+

Lanosterol

HO

CONFIGURATION
of
Lanosterol

Lanosterol is the precursor for the biosyntheses of primary metabolites that have a common structural motif of three six-membered rings and one five-membered ring. These compounds are known as **steroids**. In the chemistry of natural products, the rings are labeled by capital letters A, B, C and D.

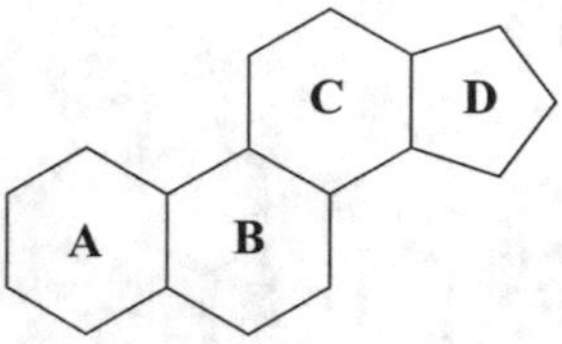

STEROID SKELETON

Some steroids such as **cholic acid**, **progesterone** and **testosterone** were already mentioned in the chapters discussing aldehydes, ketones and carboxylic acids. The most common steroid in humans is **cholesterol**. Although the compound has been discovered in XVIII century, its complete molecular structure was determined only in the middle of the last century. Cholesterol appears in most of tissues and its special role is in the regulation of blood circulation. Imbalance of cholesterol in the organism can cause serious health problems similar to artheriosclerosis. The cholesterol molecule, like other steroids, is formed by a particular biosynthetic pathway from terpene precursors, squalene and lanosterol. Since cholesterol has 27 carbon atoms, three atoms less than the triterpene squalene (which has 30 C-atoms), three C-atoms are eliminated during the biosynthetic process.

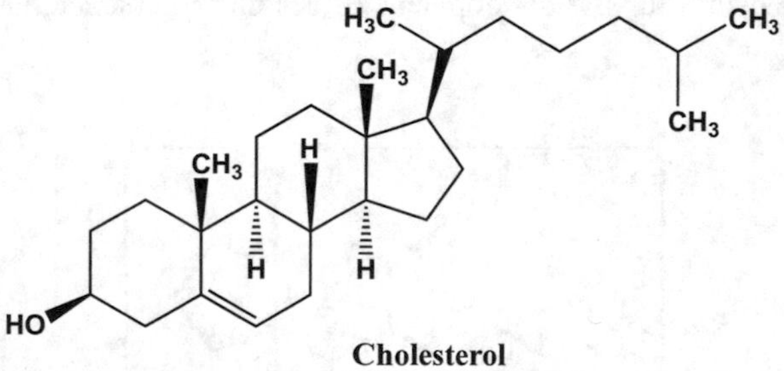

Cholesterol

Under exposure to sunlight the cholesterol in the skin is oxidized to 7-dehydrocholesterol that is immediately rearranged to **vitamin D$_3$**.

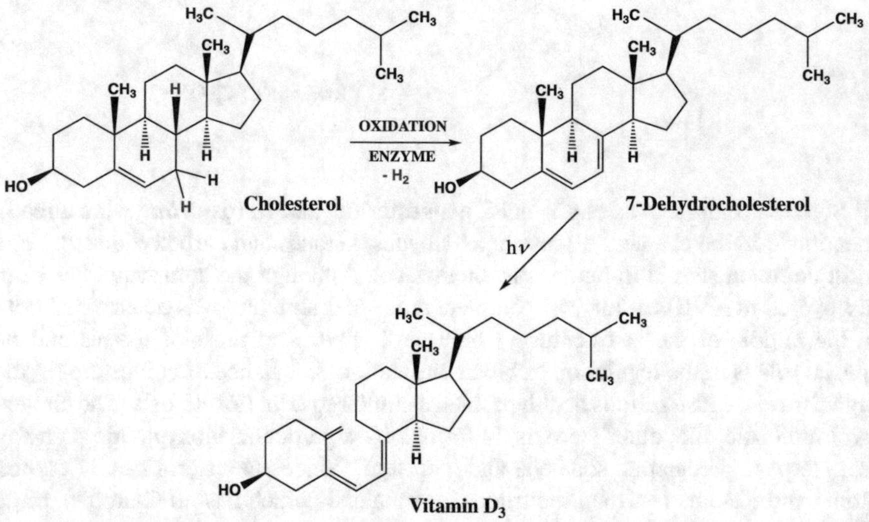

Cholesterol

7-Dehydrocholesterol

Vitamin D₃

Most hormones are also steroids. Besides the already mentioned sexual hormones testosterone and progesterone, there is also **estradiol**, the female sexual hormone, which is responsible for the development of secondary characteristics of female anatomy.

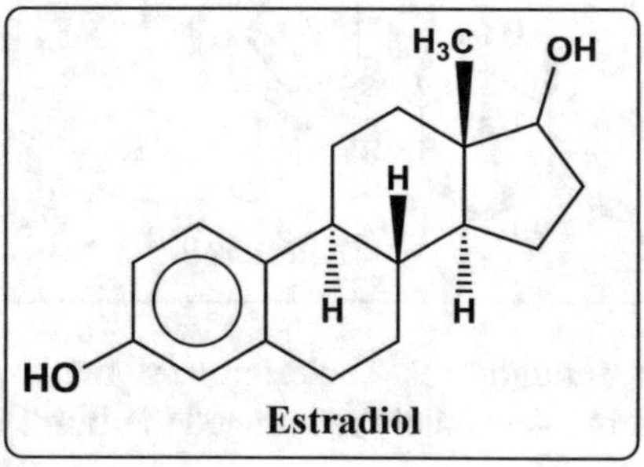

Estradiol

11.4.4.1   Terpenes as Secondary Metabolites

We have already mentioned that secondary metabolites are compounds which are specific to the particular biological species. They are not common to all living organisms. Terpenes which were discussed in the previous section are primary metabolites. However, many organisms produce secondary metabolites that are structurally similar to terpenes, i.e., their molecules can be decomposed into isoprene units. One of them is the already mentioned camphor but more examples are given below.

Mentol   α-Selinen   Cembrene   Vitamin A

Formulas of terpenes in the scheme above are presented in two rows. If in the bottom row of the scheme the chemical bonds labeled in red are deleted, the residues represent are the isoprene units. Based on the number of isoprene fragments, **menthol** is monoterpene, **α-selinene** belongs to sesquiterpenes and **cembrene** as well as **vitamin A** are diterpenes. Menthol is the component of peppermint responsible for its refreshing odor. Aroma of celery originates from **α-selinene** while **cembrene** can be isolated from pine trees. The derivative of **retinal** has already been discussed in section about the *cis–trans* isomerism in the biochemical process of vision.

11.5 Alkaloids

Secondary metabolites that have been first isolated from natural sources are alkaloids, the compounds which were the first group of natural products studied systematically by chemists. These compounds were named **alkaloids** because all of them have nitrogen in their molecules and it is known that nitrogen causes alkaline behavior. As it is typical for all secondary metabolites, the molecules of alkaloids can have a wide variety of structures. Many of alkaloids induce significant physiological effects, especially on the nervous systems. Some of them, like **nicotine** or components of opium are drugs or have analgesic properties.

Adrenaline is produced in the brain as stress activator and it causes increased psychophysical activity. Compounds isolated from special sorts of poppy have frequently been used in the form of impure substances as tranquilizers and are known under the name **opium**. The name is drawn from the Greek word οπιον which means poppy. The first pure substance isolated from opium is **heroin**, the compound that has also been synthesized for the medicinal use.

Nicotine

Adrenaline

Opimu alcaloids

R = R′ = H   Morphine

R = R′ = OCCH₃   Heroin

R = CH₃   R′ = H   Codeine

11.6   Organic and Bioorganic Reactions

Most of the reactions mentioned so far are used in laboratory preparations. The difference between bioorganic and laboratory synthesis has already been mentioned in the context of the synthesis and biosynthesis of amino acids. The Strecker synthesis was compared with reactions catalyzed by the enzyme aminotransferase. The most important difference between the laboratory synthesis and biosynthesis is that the biosynthesis is strongly stereospecific. In living cells, the reaction conditions are fundamentally different from those in the organic laboratory. In the laboratory, the reactions are performed under drastic conditions of elevated temperature or pressure, frequently by using strong reagents such as metal catalysts and/or strong acids and bases. Besides, most of the reactions in laboratory occur in organic solvents and within a wide range of acidity. In contrast, the living cell is made mostly from water and is very sensitive to strong inorganic reagents. In most cells the acidity is nearly constant with the pH value maintained close to the value of 7.5.

For the reactions such as nucleophilic substitutions, eliminations or redox reactions, living systems use not only the special catalysts such as enzymes, but also the specific biologically compatible reagents, leaving groups and nucleophiles.

As we know, the nucleophilic substitution can occur via two possible mechanisms, S_N1 or S_N2. While the S_N2 mechanism depends on the concentration and on the nature of the nucleophile, the S_N1 mechanism depends only on the nature of the leaving group. Reaction intermediates in S_N1 reactions, the carbocations, are the molecules that can be effectively stabilized by the enzyme catalysts and the reaction conditions for S_N1 processes can be milder. Consequently, most nucleophilic substitutions in living cells occur via the S_N1 mechanism. For the interconversion of functional groups by nucleophilic substitution in the laboratory we use alkyl halides as starting materials. Since alkyl halides are hydrophobic, they cannot persist in the aqueous medium of the living cell and halide as a leaving group is not compatible with bioorganic systems. Leaving groups in cells must be hydrophilic such as **pyrophosphate** esters, which are ionic molecules well soluble in water.

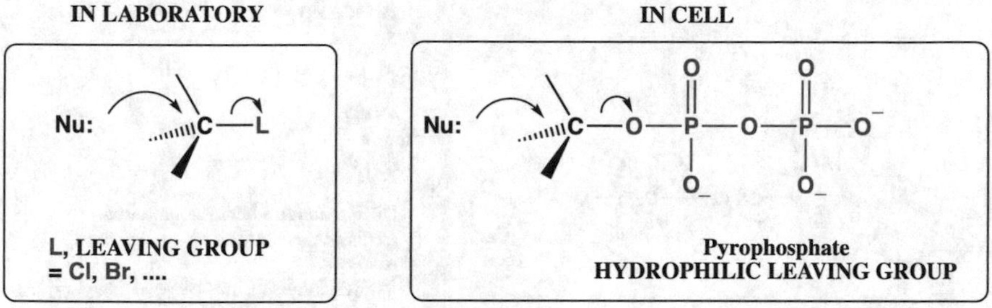

Carboxylic acid derivatives are prepared in laboratory starting from acyl chlorides. Since acyl chloride reacts vigorously with water yielding HCl, its use in living systems would kill the cell. Hence, the leaving group in the biosynthesis of derivates of carboxylic acids, the **coenzyme A** has special and more complex structure.

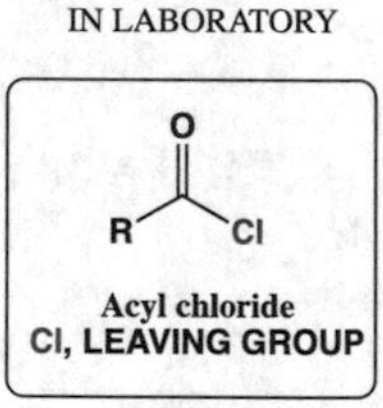

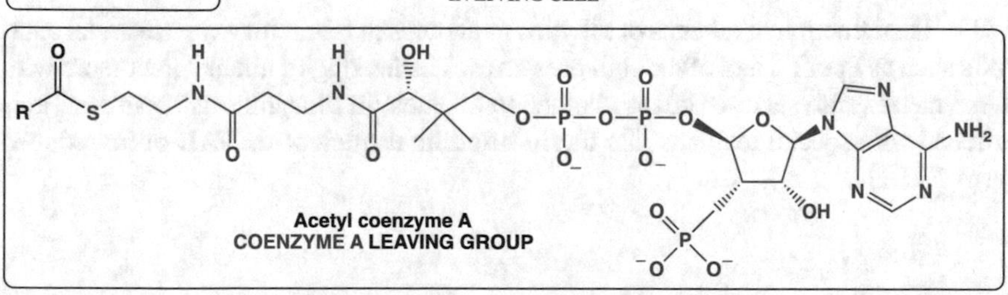

Oxidations and reductions can be performed in the laboratory by using metallates or metal hydrides. Under very mild conditions living cells use special redox systems which involve enzymes and the corresponding active reagents with complex structures.

The most common reagents are **nicotinamide adenine dinucleotide NAD+** and its phosphate derivative **NADP+**. In the reduced form these molecules are transformed to **NADH** and **NADPH** by accepting hydrogen atom.

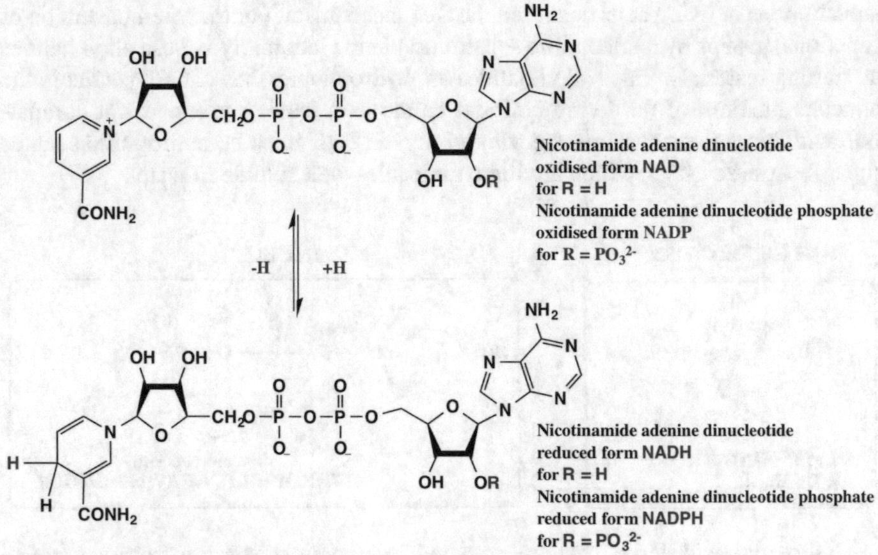

NAD$^+$ is efficient in reactions which yield functional groups with double bonds between carbon atom and the heteroatom. In this way alcohols can be converted to ketones. In laboratory we need strong oxidants such as $KMnO_4$ or $K_2Cr_2O_7$ to oxidize alcohols.

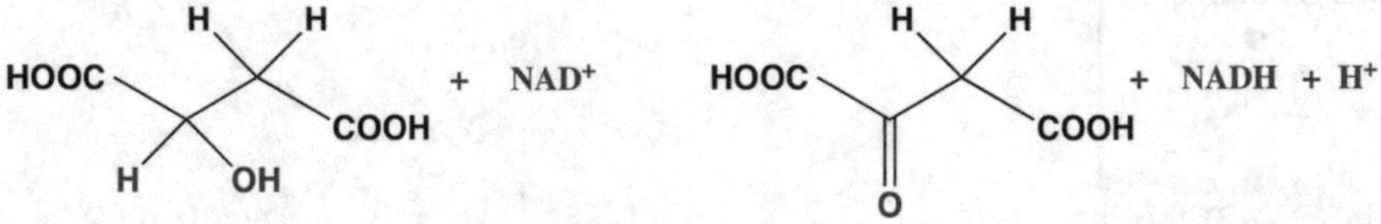

For the reduction of alkenes or alkynes to alkanes in laboratory we use metal catalysts such as Pt or Pd and often high pressures. The heating of alkane precursors with these metal catalysts re-oxidizes alkanes to alkenes. In biosynthesis these reactions proceed with special reagents like **flavin adenine dinucleotide FAD** or its reduced form **FADH$_2$**.

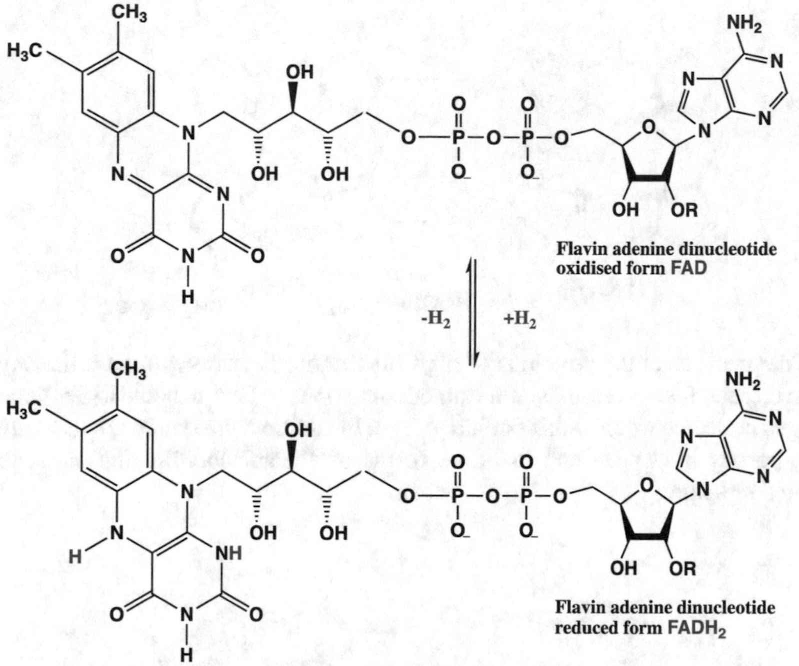

Flavin adenine dinucleotide
oxidised form FAD

-H₂   +H₂

Flavin adenine dinucleotide
reduced form FADH₂

Example of reactions mediated by FAD or FADH$_2$ is given in the scheme:

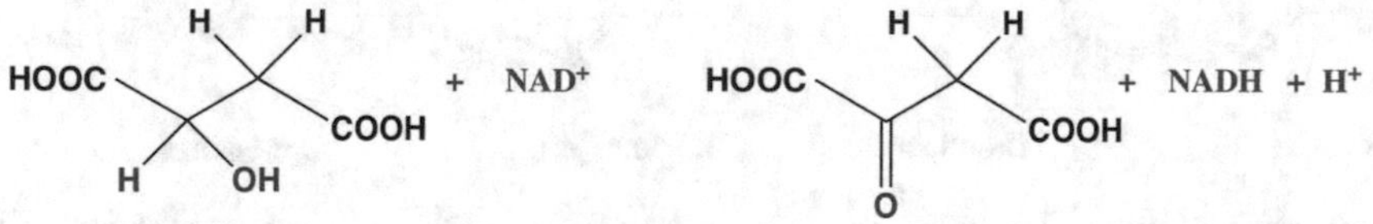

One of the most important redox processes in living organisms includes the transfer of oxygen in erythrocytes by the protein **hemoglobin**. The active functionality in hemoglobin is the organometallic group called **heme**. Organic part of this complex is **porphyrin** aromatic heterocyclic molecule (**heterocyclic structures are cyclic molecules with heteroatoms in the ring**).

The oxygen molecule binds to the iron atom situated in the molecular plane of the porphyrin ring. The opposite side of the plane is occupied by the **histidine** amino acid that is part of the polypeptide chain of the protein molecule.

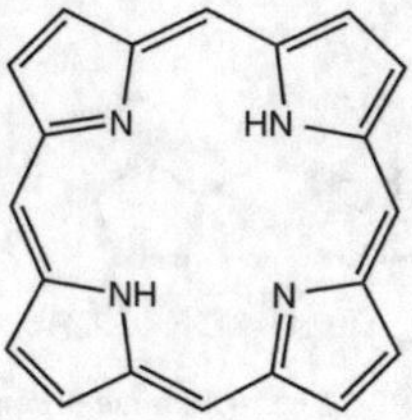

N
HN
NH
N

Porphyrin

N
HN
Fe^{2+}
NH
N
$HOOCCH_2CH_2$
CH_2CH_2COOH

Heme

The presence of the protein is of high importance because without it the oxygen atom close to the iron atom would immediately oxidize Fe^{2+} in heme to Fe^{3+} and the oxygen molecule would then not bind to iron. In addition, the iron ion in the complex with porphyrin oxygen and histidine forms octahedral coordination, a sterically favored structure.

Fe^{2+}
$HOOCCH_2CH_2$
CH_2CH_2COOH
O_2
O=O
Fe^{2+}
$HOOCCH_2CH_2$
CH_2CH_2COOH
HN
N
Histidine from protein
polypeptide chain
N
H
C=O

Heme

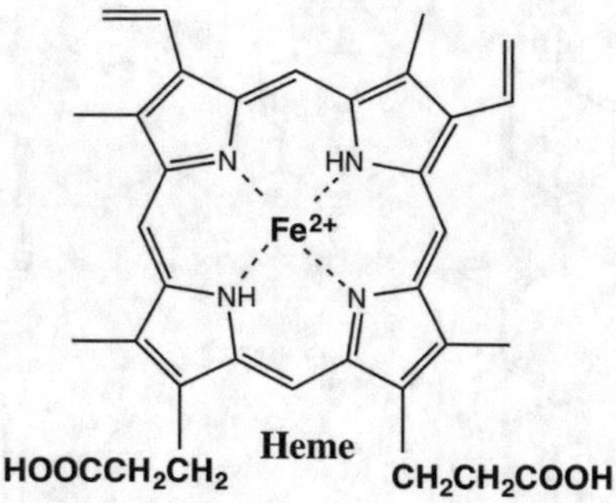

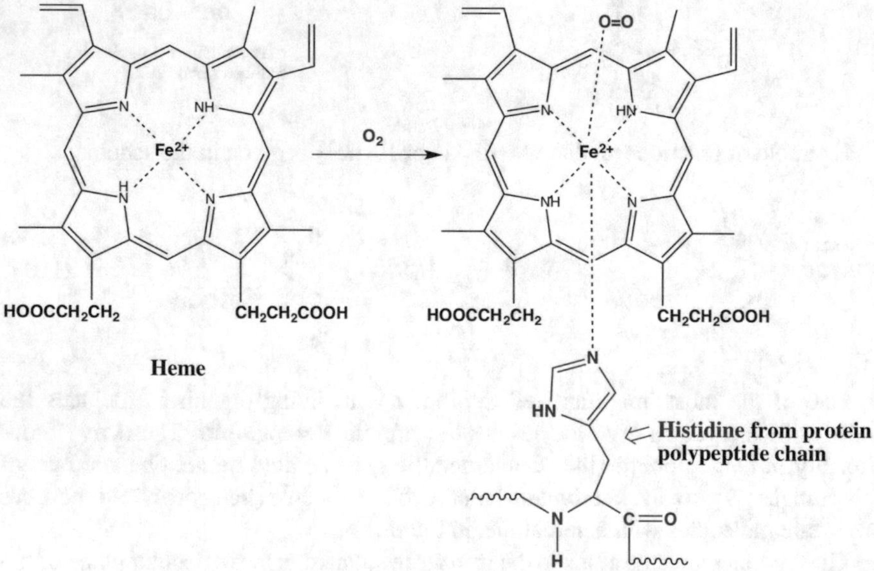

Heme - oxygen complex

In this comparison between the reactions in the living organisms and those in the laboratory, we have discussed only the most common examples. We did not mention the most important property of bioorganic reaction systems, their interconnectivity. Biochemical reactions forms organized systems that act more or less autonomously.

11.7 Organic Synthesis

As we have already mentioned, the scope of organic chemistry is focused on the structure, dynamics, and synthesis. Practical application of organic chemistry is not only in the isolation of natural products, but still more in their preparation. Human life is today highly dependent on plenty of organic substances which have their use in the design of new materials necessary in high technology, and, especially in medicine. Although pharmaceutical drugs are in some case isolated from the natural material, most of them are synthesized in laboratory and in industry.

In principle, the procedure of organic synthesis comprises a series of reactions in which the starting compound is gradually transformed in the compound of interest. Historically, organic synthesis has been developed in the second half of XIX century, when some of basic reaction types were discovered. The research of the most of organic chemists has been focused on the synthesis of natural products. Their composition and the structure were then compared with the properties of the same compounds isolated from the natural sources. This approach has been for long period of history of organic chemistry one of the main methods for confirmation of the molecular structure of natural compounds.

The procedure of organic synthesis is the design, the invention in which the reactants are transformed through successive series of reactions in the desired product. One example of such procedure is the synthesis of α-terpenoid *limonene*, which has been invented by William Henry Perkin in 1904 (Fig. 11.1).

Perkin started his synthesis by condensation of two reactants, functionalized esters. In the second step the compound **A** is transformed to **B** by oxidation of nitrile group to carboxylic acid and successive elimination of ester. In the next step the structure **C** is obtained by decarboxylative cyclization. The ester **D** was prepared by classical esterification under acidic conditions. By using the Grgnard reaction, the methyl group is added to the carbonyl carbon atom (**E**). Bromide **F** was prepared by classical nucleophilic substitution, and transformed to **G** by elimination of HBr. The ethyl ester **H** was prepared by a standard method of esterification. Product **I** was obtained by the Grignard addition of methyl group on the ester function. Finally, the elimination of water molecule yields the desired product limonene (**J**).

The unavoidable question is how the synthetic procedure can be invented? Is it only a series of random attempts, or there is some systematic method that could guide chemist to find the best way that will end with the compound of interest?

The winner of Nobel Prize for the year 1990, **Elias James Corey** has invented the original method for designing organic synthesis of complicated organic molecules. The method is known as **retrosynthetic analysis**. The target compound is theoretically decomposed to fragments which correspond to the possible molecules from which this final compound could be prepared by known reaction types. These fragments can be successively decomposed to sub-fragments till the structure of the reactant that is easily available.

Let us demonstrate this procedure on the example of the preparation of relatively simple organic compound **E**:

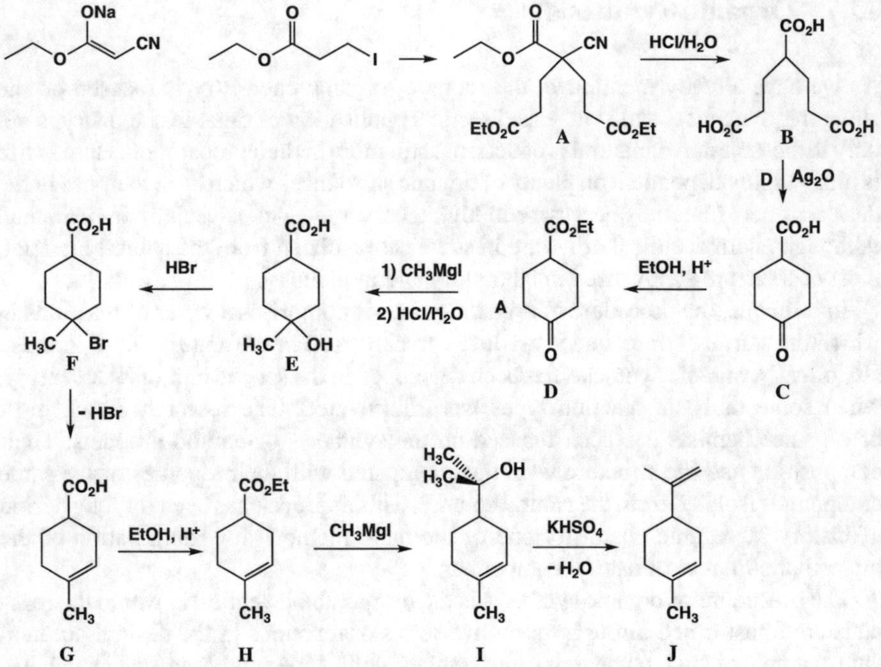

Fig. 11.1 Synthesis of limonene by W. H. Perkin in 1904

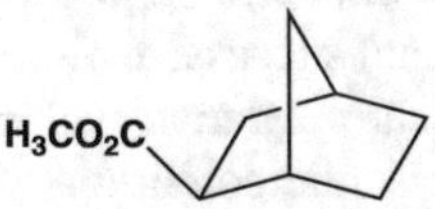

E

By analysis its structure we can recognize the norbornyl bicyclic skeleton with the methyl ester functional group. The retrosynthetic pathway should in principle be based on the decomposition of ester and on the decomposition of norbornyl skeleton in simpler fragments. Such retrosinthetic way is demonstrated in Fig. 11.2.

The final ester **E** can be prepared by standard esterification of **D** with methanol under acidic conditions. Since **D** is carboxylic acid, it could easily be obtained through functional group transformation, the oxidation of nitrile **C**, which can be prepared by nucleophilic substitution of bromide **B** with CN⁻ nucleophile. **B** can be easily formed by addition of HBr on the norbornene **C**. Deconstruction of this bicyclic structure shows that **A** is the product of Diels–Alder $4\pi + 2\pi$ addition.

Based on such retrosynthetic deconstruction, it is possible to propose the synthesis of **E** as it is shown in Fig. 11.3.

Synthesis of complex organic molecules can be a long-term scientific work, and it requires the knowledge of enormous number of chemical reactions and their practical performances in laboratory. A big help in planning organic synthesis is the method of computer-guided design of retro-synthetic decompositions and proposal of the

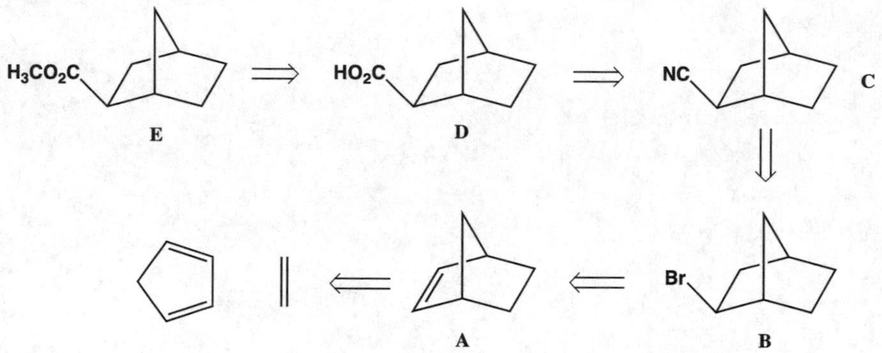

Fig. 11.2 The example of retro-synthetic deconstruction of compound **E**

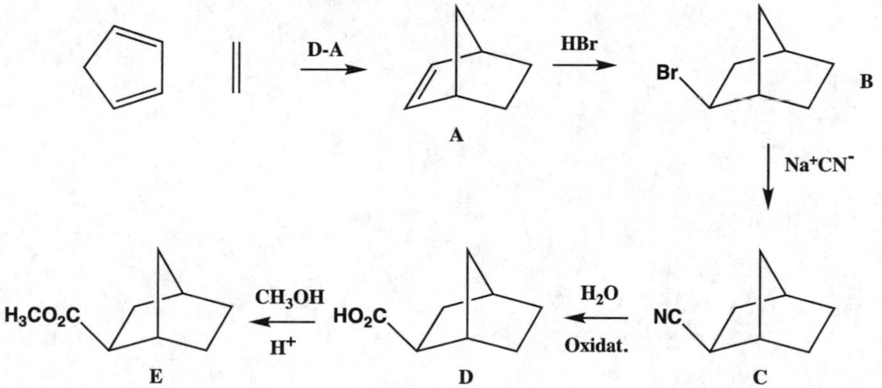

Fig. 11.3 The proposed synthesis of compound **E**

synthetic pathway. In computer systems it is possible to collect large data bases of organic transformations. However, the synthetic work in laboratory is full of surprises, and the human creative activity is, and it will be the crucial factor.

Chapter 12
Organic Supramolecular and Supermolecular Structures

Simple organic molecules can be combined to form the large molecular structures. Interconnections of such molecular subunits, the **building blocks**, can be based on the weak interactions. In this case we are talking about **supramolecular** structures. On the other hand, **supermolecules** appear if the molecular building blocks are bound by covalent interactions. The simple supramolecular structures are crown ethers, which have already represented in Sect. 5.2.3. Here we wish to discuss some of the basic principles responsible for the formation of super- and supramolecular structures.

Analyzing the structures of the molecules of organic compounds, which have been discussed in previous chapters, we could recognize that these structures are composed from some basic structural patterns. First of all, there are characteristic spatial configurations, linear, trigonal, and tetrahedral. In both, 2D (trigonal), and 3D (tetrahedral) structural arrays the **fundamental geometrical parameters are angles between the chemical bonds**. Under these angles, chemical bonds are directed in space. In the Table 12.1 are listed the basic structural angles together with the representative simple molecules. Characteristic directionality of chemical bonds enables that two or more molecules approaching to each other by moving through space can, or cannot form the complex by **weak interactions**, or the supermolecule by covalent bonds. In other words, the molecules approaching to each other can, or cannot **recognize** each other. The molecules can recognize each other if they have the complementary structural pattern based on the orientations of interactive functional groups, which are determined by the angles between chemical bonds. Such principle of the molecular recognition is illustrated in the Fig. 12.1.

Such molecules, which are programed to be compatible with each other in spatial orientations because of the characteristic steric structures determined by inter-bond angles, become in such a way self-assembled by weak intermolecular interactions.

Weak intermolecular interactions have significantly lower binding energy in comparison with covalent bonds. The most important weak interactions are hydrogen bonding, charge to charge interactions, donor–acceptor complexing, π-π interactionsπ-π, van der Waals attractions, etc. (Fig. 12.2).

© The Author(s), under exclusive license to Springer Nature Switzerland AG 2022
H. Vančik, *Basic Organic Chemistry for the Life Sciences*,
https://doi.org/10.1007/978-3-030-92438-6_12

Table 12.1 Typical examples representing the fundamental angles between the "outer" chemical bonds (labelled with bold lines) of organic molecules which are responsible for association of molecules in supramolecular structures or supermolecules

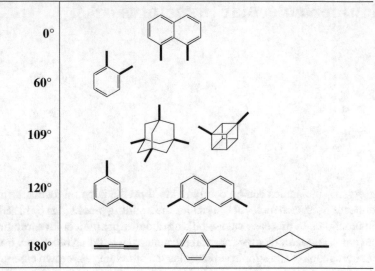

Fig. 12.1 Molecular recognition: complexing of two simple amide molecules (**orientation 0° in Table 12.1**)

As we have already mentioned, the molecular aggregates-complexes which are formed by molecular recognition are called supramolecular structures which are self-assembled from the building blocks, i.e., the smaller molecular units which interact by weak interactions. The most important fact is that supramolecules possesses **new properties**, which are more than the sum of the properties of their precursor building blocks. In the nature (in living organisms), supramolecules could be combined with supermolecules, the natural polymers such as proteins, polysaccharides, and polynucleotides.

As it is represented in the Table 12.1, there are typical orientation angles (0°, 60°, 109°, 120°, and 180°) of the "outer" chemical bonds which are connectors of one with another molecular building block. Examples of the supramolecular structures that are constructed from building blocks with different orientation angles are represented in a series of the following figures. One of the most frequent orientations of molecules in

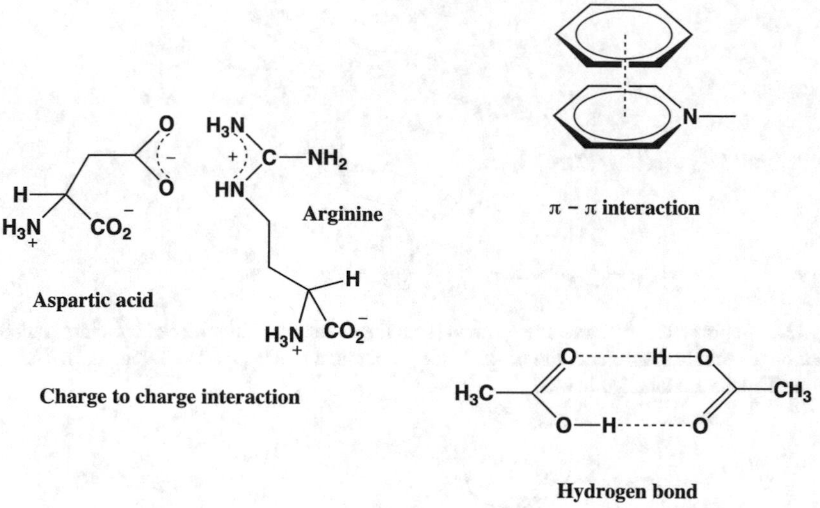

Fig. 12.2 Examples of weak interactions between molecules

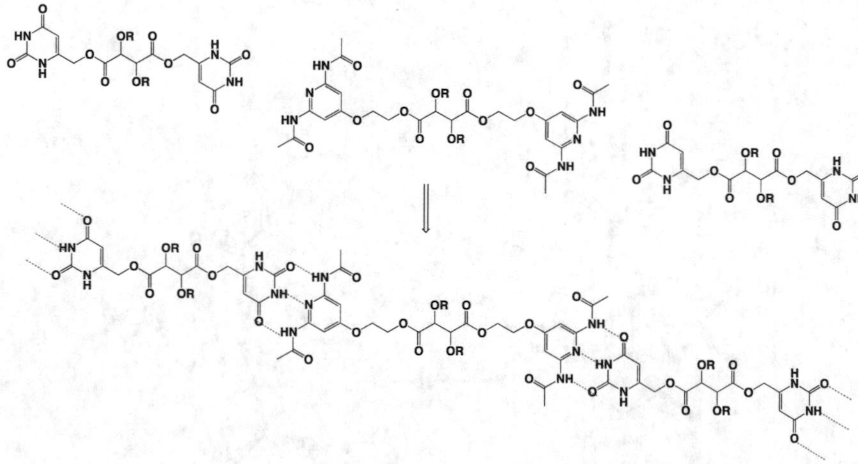

Fig. 12.3 Supramolecular structures formed from the building blocks by hydrogen bonds in parallel orientation (0° in Table 12.1). Hydrogen bonds are labelled with red dotted lines

living nature is parallel orientation of hydrogen bonds that typically connect hydrogen from NH, or OH group with the lone pair of amino or carbonyl group (Fig. 12.3).

In the interesting polycyclic molecule, **cubane**, the outer chemical bonds on the diagonal carbon atoms are oriented under the tetrahedral angle of 109°. This orientation enables dimerization of two cubanecarboxylic acids (Fig. 12.4). The most useful building block with the tetrahedral orientation is the adamantane molecule in which

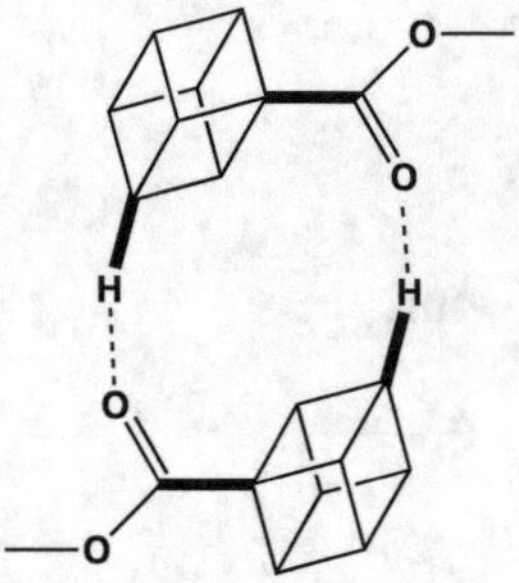

Fig. 12.4 Supramolecular structures formed from the cubane building blocks by hydrogen bonds based on the tetrahedral orientation of the "outer" chemical bonds (109° in Table 12.1). The outer chemical bonds are labelled in bold

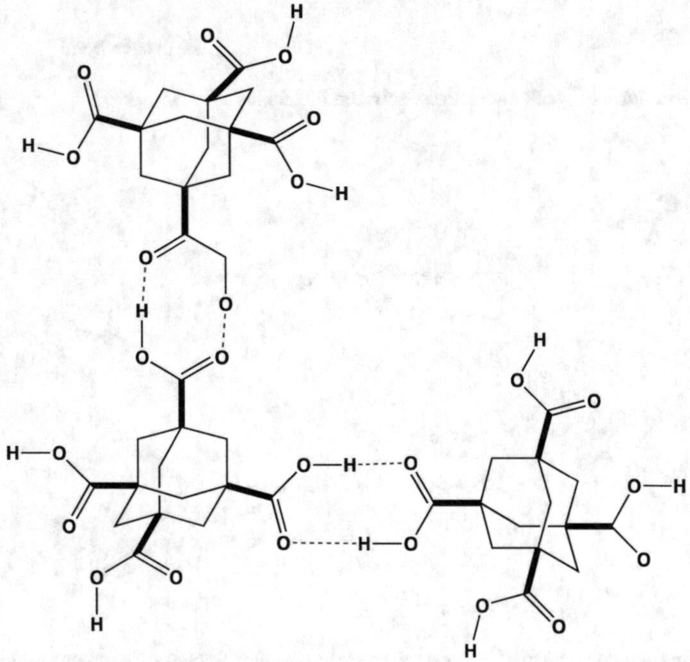

Fig. 12.5 Supramolecular structures formed from the adamantane building blocks by hydrogen bonds based on the tetrahedral orientation of the "outer" chemical bonds (109° in Table 12.1). The outer chemical bonds are labelled in bold

the bonds on carbon atoms on the positions 1, 3, 5, and 7 are oriented under 109° (Fig. 12.5).

Intermolecular assembly can be obtained also by the formation of weak bonds either between two halogen atoms, or between halogen and another heteroatom. In Fig. 12.6 is represented formation of the supramolecular structure based on the weak

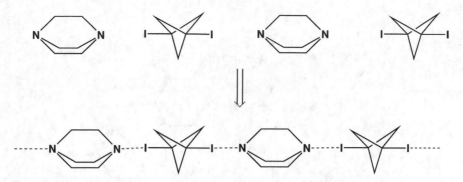

Fig. 12.6 Supramolecular structures formed from the building blocks by halogen-nitrogen bonds based on the linear orientation of the "outer" chemical bonds (180° in Table 12.1)

iodine-nitrogen bond in linear orientation. This is an example for the orientation angle 180°.

The planar building block molecules with aromatic ring, and the orientation of outer chemical bonds at 120° can form the 2D supramolecular structures (Fig. 12.7).

12.1   Molecular Recognition and Catalysis

Intermolecular recognition is the principle that *inter alia* explains the organic reactions enhanced by catalysis and autocatalysis. In the autocatalytic process the product of the reaction serves as a catalyst. It can be demonstrated by following equations:

$$A + B \rightarrow C \tag{12.1}$$

$$A + B + C \rightarrow 2C \tag{12.2}$$

In the Fig. 12.8 is shown the example of catalytic-autocatalytic reaction that is the consequence of molecular recognition and supramolecular assembly of two reactants with the molecule that is in the same time the product of their condensation. Molecule of the product, C, has functional groups is such stereochemical arrangement, that two molecules A and B, which are structured for the weak interaction, can form the complex. Such characteristic arrangement of functionalities on the molecule C, are **molecular receptors**. The reaction in Fig. 12.8 represents the general principle of the mechanism by which the complex chemical systems in living organisms occur.

Receptors can have a variety of structures, from simple molecular matrix like C, to the complicated spatial arrangements of functional groups in the active site of the enzyme catalyst. In the enzyme active site, the functional groups responsible for the interactions of enzyme molecule with the substrate by weak interactions are in most cases the residues of amino acids, the components of polypeptide chains. Active site

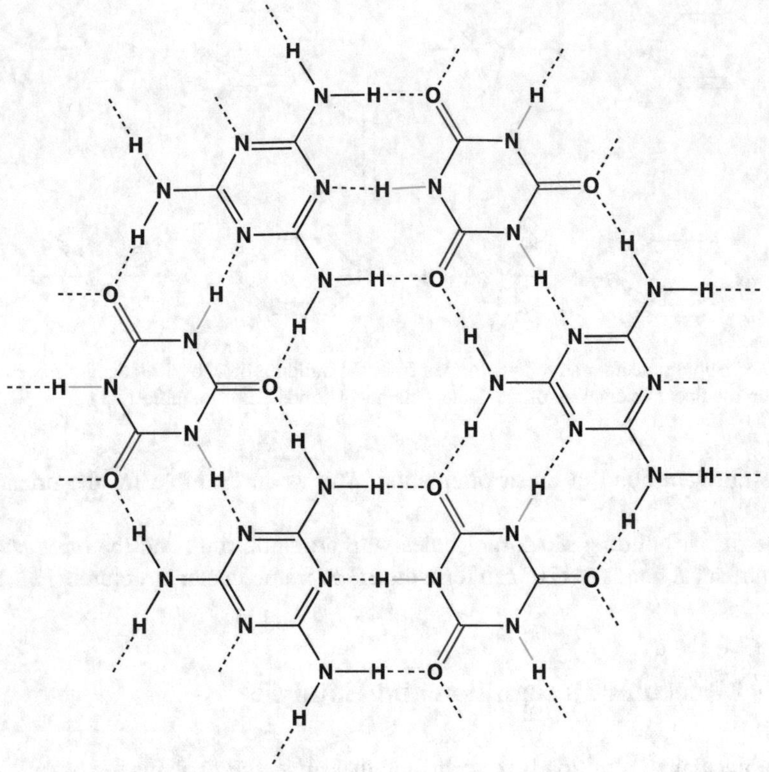

Fig. 12.7 Supramolecular structures formed from the building blocks by hydrogen bonds based on the trigonal orientation of the "outer" chemical bonds (120° in Table 12.1). The two sets of outer chemical bonds are labelled in red, and in green

is the **cavity** in the enzyme molecule, which by molecular recognition selects only those substrate molecules which have the shape and the configuration of functional groups complemental to the arrangement of functionalities in this cavity. (Fig. 12.9).

Molecular recognition between enzyme active site and the corresponding organic molecule is the basic principle how the chemical reactions in the living systems work. Most of the reactions in organisms are catalyzed and regulated in accordance with such principle of molecular recognition between enzyme and substrate.

As an illustrative example of the enzyme–substrate interaction let us talk about a group of enzyme molecules which possess characteristic active site that can catalyze the reaction of nucleophile addition to the carbonyl group. These active sites consist from the three amino acid residues, of which one is acidic, one basic, and one serves as nucleophile. Such combination of three amino acids in the enzyme cavity is called the **triad**. Their arrangement in the enzyme molecule, as well as their reaction mechanism of the nucleophile attack is shown in Fig. 12.10.

However, the enzyme molecule can recognize also molecules which have corresponding configuration of the functional groups, but which are not the substrates for

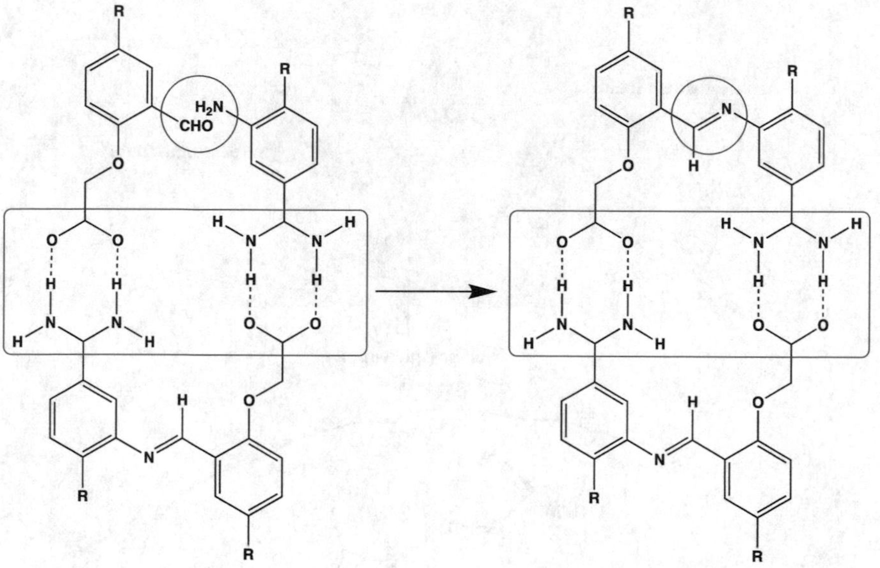

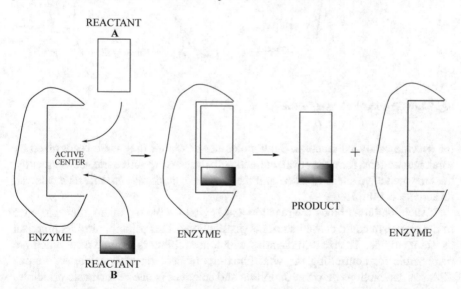

Fig. 12.8 Autocatalytic reaction based on the molecular recognition between the reactants and the product of their condensation. Molecular receptors are within the blue frame

Fig. 12.9 Molecular recognition in the active site (active center, cavity) of the enzyme molecule, and the general mechanism of the enzyme catalyzed reaction

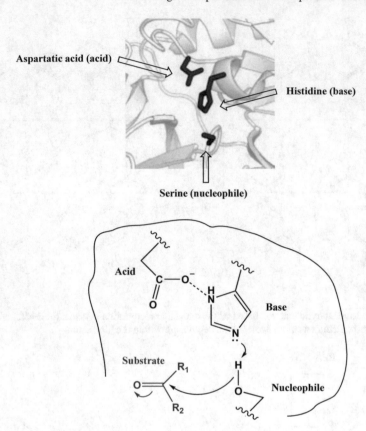

Fig. 12.10 Triad in the enzyme active center

the specific catalyzed reaction. Such molecules can, after they were **recognized** by weak interactions, form the covalent bond with the enzyme active site. Consequently, the enzyme active site is blocked and the enzyme cannot play its role as a catalyst, its activity is **inhibited**.

If such inhibitors block the reaction that is vital for the organism, their presence in the organism could cause its critical disfunctions. This principle has its practical use in medicine. There are molecules which are inhibitors of enzymes which are responsible for controlling the vital functions of bacteria responsible for human diseases. Interactions between inhibitors and enzymes is one of the basic principles of the medical drug design.

One of the first drugs with such antibiotic activity has been discovered 1928. **Alexander Fleming** (1881. 1955) has isolated the substance with the high antibiotic activity from the species *Penicillium notatum*, and he named it **penicillin**. The molecule consists from two condensed rings of which the reactive functional group is the four-membered β-lactam ring (Fig. 12.11). About 10 types of penicillins which

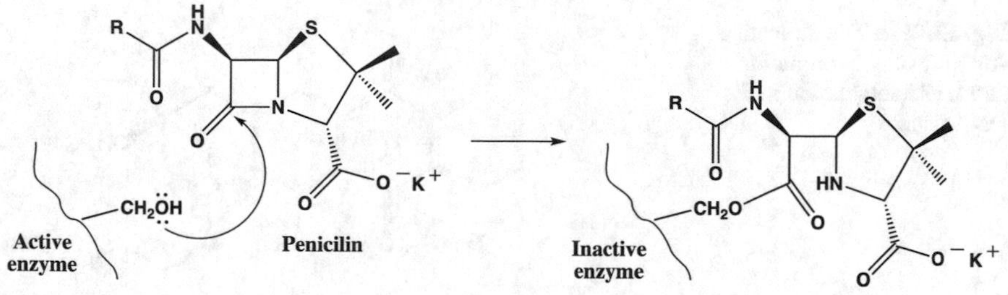

Fig. 12.11 Inhibition of the enzyme activity by reaction with penicillin

are in clinical use differ in the group R. Since for the medical use as a drug, peni-cillin should be soluble in water, it is converted to the potassium salt (Fig. 12.11). As we have discussed in previous chapters, four-membered rings are strained structures with high potential energy. Consequently, such molecules are reactive because they intend to release the strain energy. This is the reason that the molecule of penicillin reacts readily with the -CH_2OH of the enzyme active site by forming the ester group. The activity of the enzyme molecule is in such a way inhibited because its active site is blocked. Since this enzyme is responsible for the synthesis of the bacterial membrane walls, its inhibition kills bacteria because they cannot more prepare their cellular membranes. The mechanism of this reaction is described as a nucleophilic addition of alcohol to the carbonyl group in the lactam ring (Fig. 12.11).

The mechanism of drug activity for the specific diseases is based also on the weak interaction between drug molecule and DNA. As it is discussed earlier in this book (Sect. 11.3), nucleic base pairs in the double stranded DNA are arranged by the parallel position of the planar rings. The distance between these rings of base pairs is 340 pm (3.4 Å). Because of the flexibility of the DNA polymer molecule, it is possible that other planar molecules penetrate between these base pairs and form the complex by π-π interactionsπ-π. The example of such interaction is the **intercalation** of the **daunorubicin** molecule in DNA (Fig. 12.12a and b).

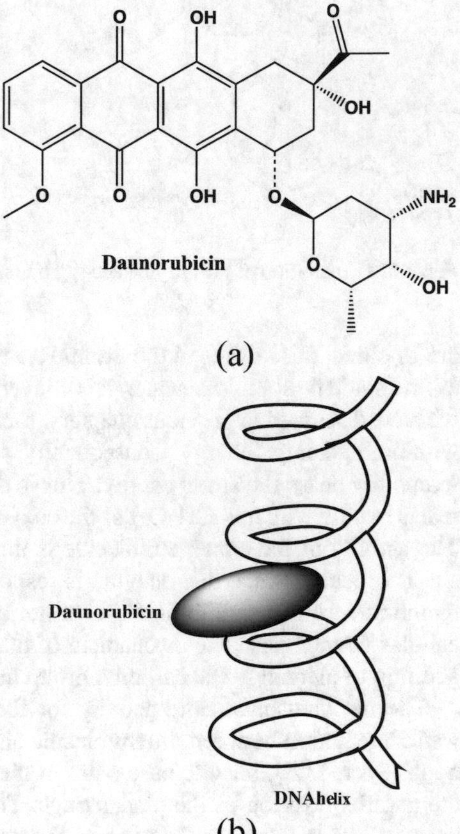

Fig. 12.12 a The molecular
structure of daunorubicin,
and **b** its intercalation in the
DNA helix

Daunorubicin

(a)

Daunorubicin

DNAhelix

(b)

The replication of the DNA molecule is in such complex is completely blocked.
Substances whose molecules have such planar rings that can intercalate in DNA are
used either as cytostatic agents in cancer therapy, antiviral drugs, or as antibiotics.

Index

A

Absolute configuration, 114, 147
(*R*) absolute configuration, 115
(*S*) absolutna configuration, 115
Absolute methanol, 76
Acetal, 94, 95
Acetaldehyde, 93
Acetamide, 124
Acetanhydride, 118
Acetic acid, *acidum aceticum*, 102, 103,
   118, 119, 124
Acetone, 35, 91, 100
Acetylene, 43
Acid halides, 117
Acids, 54, 77–81, 87, 96, 97, 100–105, 109,
   118–122, 127, 130, 147–149, 158,
   169–171, 178, 189
Acidum aceticum, 102
Acidum formicum, 102
Acrolein, 92
Activation energy, 71, 72, 74
Activation (of benzene ring), 131
Active center (of enzyme), 193, 194
Acylation, 129
Acyl group, 117, 129
Adamantane, 18, 189, 190
Addition, 1, 50–57, 60, 63, 73, 77, 88, 90,
   100, 101, 122, 127, 128, 138, 142,
   157, 161, 170, 171, 182–184, 192
Adenine, 165, 167, 168
Adenosine, 165, 166
Adenosine diphosphate (ADP), 166
Adenosine monophosphate (AMP), 166
Adenosine triphosphate (ATP), 166
Adipic acid, 124
Adrenaline, 177

Alanine, Ala, 147
Alcoholates, 76, 78, 81, 94
Alcohols, 20, 34, 69, 74–86, 90, 94, 97, 99,
   100, 105, 118–122, 161, 169, 170,
   180, 195
Aldehyde, 20, 34, 56, 90–101, 118, 121,
   147, 155–159, 175
Alder, Kurt, 137
Aldohexose, 155, 156
Aldol, 97, 98
Aldol condensation, 97, 98
Aldopentose, 156, 158
Aldose, 155, 158, 159
Aldotriose, 157
Alkaloids, 141, 177
Alkanes, 1, 3–6, 8–11, 14, 19, 34, 43, 44,
   50, 124, 171, 180
Alkenes, 1, 34, 43–46, 48–57, 60, 61, 80,
   81, 114, 127, 137, 138, 151, 170,
   171, 173, 180
Alkoxides, 76, 81, 117
Alkylation, 129
Alkyl halides, 51, 63, 64, 67, 69, 81, 82, 85,
   99, 101, 129, 179
Alkynes, 1, 34, 43–45, 51, 54, 90, 170, 180
Allene, 44
Amides, 35, 117, 121, 123–125, 151, 188
Amines, 35, 85–87, 99, 117, 122–124, 165
Amino acids, 141–144, 146–151, 154–156,
   178, 181, 191, 192
α-Amino acids, 142
β- Amino acids, 142
γ-Amino acids, 142
Aminotransferase, 148, 178
Ammonia, 24, 85–87, 123
Ammonium cyanate, 2, 3

© The Editor(s) (if applicable) and The Author(s), under exclusive license
to Springer Nature Switzerland AG 2022
H. Vančik, *Basic Organic Chemistry for the Life Sciences*,
https://doi.org/10.1007/978-3-030-92438-6

Printed in the United States
by Baker & Taylor Publisher Services